# The Epistemology of Development, Evolution, and Genetics

Collected for the first time in a single volume are essays that examine the developments in three fundamental biological disciplines – embryology, evolutionary biology, and genetics – which were in conflict for much of the twentieth century. The essays in this collection examine key methodological problems within these disciplines and the difficulties faced in overcoming the conflicts between them. Burian skillfully weaves together historical appreciation of the settings within which scientists work, substantial knowledge of the biological problems at stake, and the methodological and philosophical issues faced in integrating biological knowledge drawn from disparate sources. The final chapter describes what recent findings in developmental biology and genetics can tell us about the history and development of animals.

Written in a clear, accessible style, this collection should appeal to students and professionals in the philosophy of science and the philosophy and history of biology.

Richard M. Burian is Professor of Philosophy and Science Studies at Virginia Polytechnic Institute and State University.

CAMBRIDGE STUDIES IN PHILOSOPHY AND BIOLOGY



Alfred I. Tauber *The Immune Self: Theory or Metaphor?*
Elliott Sober *From a Biological Point of View*
Robert Brandon *Concepts and Methods in Evolutionary Biology*
Peter Godfrey-Smith *Complexity and the Function of Mind in Nature*
William A. Rottschaefer *The Biology and Psychology of Moral Agency*
Sahotra Sarkar *Genetics and Reductionism*
Jean Gayon *Darwinism's Struggle for Survival*
Jane Maienschein and Michael Ruse (eds.) *Biology and the Foundation of Ethics*
Jack Wilson *Biological Individuality*
Richard Creath and Jane Maienschein (eds.) *Biology and Epistemology*
Alexander Rosenberg *Darwinism in Philosophy, Social Science, and Policy*
Peter Beurton, Raphael Falk, and Hans-Jörg Rheinberger (eds.) *The Concept of the Gene in Development and Evolution*
David Hull *Science and Selection*
James G. Lennox *Aristotle's Philosophy of Biology*
Marc Ereshefsky *The Poverty of the Linnaean Hierarchy*
Kim Sterelny *The Evolution of Agency and Other Essays*
William S. Cooper *The Evolution of Reason*
Peter McLaughlin *What Functions Explain*
Steven Hecht Orzack and Elliott Sober (eds.) *Adaptationism and Optimality*
Bryan G. Norton *Searching for Sustainability*
Sandra D. Mitchell *Biological Complexity and Integrative Pluralism*
Gregory J. Cooper *The Science of the Struggle for Existence*
Joseph LaPorte *Natural Kinds and Conceptual Change*
William F. Harms *Information and Meaning in Evolutionary Processes*
Markkn Oksanen and Juhani Pietarinen *Philosophy and Biodiversity*

# The Epistemology of Development, Evolution, and Genetics

## Selected Essays

RICHARD M. BURIAN

*Virginia Polytechnic Institute and State University*

PUBLISHED BY THE PRESS SYNDICATE OF THE UNIVERSITY OF CAMBRIDGE
The Pitt Building, Trumpington Street, Cambridge, United Kingdom

CAMBRIDGE UNIVERSITY PRESS
The Edinburgh Building, Cambridge CB2 2RU, UK
40 West 20th Street, New York, NY 10011-4211, USA
477 Williamstown Road, Port Melbourne, VIC 3207, Australia
Ruiz de Alarcón 13, 28014 Madrid, Spain
Dock House, The Waterfront, Cape Town 8001, South Africa

http://www.cambridge.org



First published 2005

Printed in the United States of America

*Typeface* Times Roman 10.25/13 pt. *System* LaTeX $2_\varepsilon$ [TB]

*A catalog record for this book is available from the British Library.*

*Library of Congress Cataloging in Publication Data*

Burian, Richard M.

The epistemology of development, evolution, and genetics : selected essays / Richard Burian.

p. cm. – (Cambridge studies in philosophy and biology)

Includes bibliographical references and index.

ISBN 0-521-83675-1 (hbk.) – ISBN 0-521-54528-5 (pbk.)

1. Developmental biology – Philosophy. 2. Evolution (Biology) – Philosophy. 3. Genetics – Philosophy. 4. Knowledge, Theory of. I. Title. II. Series.

QH491.B86 2005
571.8 – dc22 2004044242

ISBN 0 521 83675 1 hardback
ISBN 0 521 54528 5 paperback

*For Anne:*
*Companion, friend, constant source of inspiration;*

*and for three people without whom this book would not exist:*
*Jean Gayon, Marjorie Grene, and Michael Ruse*

# Contents

# Preface

Forty-one years ago, in August 1963, I entered graduate school in the Department of Philosophy at the University of Pittsburgh. At that time, the "post-positivist" reaction to logical empiricism in philosophy of science was well under way, although not yet in full swing. In evolutionary biology, the "synthetic theory of evolution" seemed to have won a clear victory over its major rivals. In molecular biology, Watson and Crick's proposed structure for DNA, just a decade old, was widely accepted and was serving as a fulcrum for restructuring genetics as part of molecular biology, even though "the" genetic code had not yet been fully solved. (I use scare quotes because, within twenty years after the code was fully deciphered in 1966, it was discovered that some organisms have small systematic variations in the correlations between codons and amino acids. The code can vary even within a single organism: The mitochondria of many organisms employ slight variations on the standard code so that the nuclear and mitochondrial codes differ in specific ways.)

As a physics major and then as a math major in college, I had taken no courses in biology. But, my father was a research ophthalmologist and, within two years of starting graduate studies, I realized that the sort of work he did (which I knew largely by osmosis) did not fit well with the dominant positions in philosophy of science. The then-current philosophical attempts at linguistic and structural analyses of confirmation, explanation, falsification, and "rational reconstruction" of theory-based reasoning seemed of little value in providing accounts of the work that, on minimal knowledge, I took to be central within much of biology. By the time I left Pittsburgh, I had resolved, when possible, to spend a year studying one or two branches of biology. Ideally, the disciplines in question should have strong theoretical components, yet they should involve both field observation and laboratory experiments. Thanks to a generous study fellowship from the American Council of Learned Societies and the sponsorship of Stephen Jay Gould and Ernst Mayr, I was able to do just that in 1976–7, studying mainly evolution, genetics, and their histories at the Museum of Comparative Zoology and the Department of History of Science at Harvard University. Mayr was on leave for half the year, so I worked mainly with Gould and Richard Lewontin, taking courses not only with both and participating in their lunch seminars, but also with E. O. Wilson, Fotis Kafatos (a developmental biologist), Mark Ptashne (a molecular geneticist working on phage λ), and a number of other leading figures in allied fields. This was a major turning point in my academic career. I was fascinated by the biology I studied, the disputes among the leading figures in numerous biological disciplines, the variety of the experimental and fieldwork brought to bear to support or to test hypotheses, and the extent to which resolution of disputes depended on appropriate technology (including, in many cases, suitably "domesticated" laboratory organisms appropriate to the problem at hand). I was struck by the extraordinary difficulty of giving an adequate account of explanations that crossed (ill-defined) "levels" – molecular versus cellular, organismal versus populational, ecological versus evolutionary, and the like.

It is fitting to acknowledge here the transformative influence of Marjorie Grene on my work in history and the philosophy of biology. All but two of my articles in these domains were begun after I sat in on a course on evolutionary biology that she taught at Temple University in 1979. Shortly afterward, she asked me to assist her in running a summer institute on teaching the philosophy of biology held at Cornell University in 1982 under the auspices of the Council for Philosophical Studies. A group of us from that institute, joined by many others in the Autumn of 1982, organized a series of informal meetings that evolved into the International Society for History, Philosophy, and Social Studies of Biology (known as Ishkabibble, or just ISH, to its friends). The spirit of cooperative interdisciplinarity of that summer institute, also embodied in the meetings of Ishkabibble, helped to shape my work. I hope that the chapters of this book convey some of the sense of community we had then and that they reflect how much I learned from working in collaboration with others.

Indeed, I have been fortunate to work with historians and philosophers interested in biological topics and with biologists working on development, evolution, genetics, and the theoretical biology of complex systems. Due to the defects of my memory, I am often unable to disentangle the extent to which my work draws on the contributions of many friends, collaborators, critics, and students. I apologize here to all of those whose contributions I fail to mention, for I owe an enormous intellectual debt to many individuals beyond those named in this preface. I am also glad to acknowledge, with gratitude, the support of my colleagues in Philosophy and in Science and Technology Studies at Virginia Tech. Perhaps the most influential among the many people who have helped me rethink my views along the way are the three Gs: Jean Gayon, Scott Gilbert, and Marjorie Grene. Their influence has been matched (and tempered) by Ron Amundson, Robert Brandon, Lindley Darden, Michael Dietrich, Stuart Kaufman, Richard Lewontin, Jane Maienschein, Anne McNabb, Robert Richardson, Denis Thieffry, William Wimsatt, Doris Zallen, and many others. I am grateful to these dear friends and to a host of others. I am also grateful to Andrew Garnar for his help with many matters connected with this book, especially for his help in preparing the illustrations.

Finally, and most important, Anne McNabb has been a constant friend and companion for twenty years. She not only has been enormously supportive but also has been always ready with a challenge whenever I grew complacent. To her, I owe far more than I can ever say or repay.

# 1

# General Introduction

Unlike many natural philosophers of the seventeenth century, whose work bridged what we now call science and philosophy, most twentieth-century philosophers of science did not undertake serious work in what we would now call science. Often enough, they wrote in reaction to earlier writings of scientists and philosophers of science, drawing on general background knowledge. Although the theories and analyses they produced were often fascinating, this narrow way of working put them at risk of making poor contact with the phenomena of concern to scientists. Some philosophical projects fell victim to this risk, including some attempts to construct general theories of scientific method, to develop criteria for distinguishing living from nonliving entities, and to specify the structure of major biological theories. Often enough, there turned out to be more things in heaven and earth than were dreamt of in our philosophies (Shakespeare, *Hamlet*, I. v). Philosophy, including philosophy of science, should begin with wonder at the phenomena that require understanding.

The development of the philosophy of biology in the last thirty years or so has been salutary in this regard as it has become ever more involved with biological phenomena. Like many contemporary philosophers of biology, I maintain that the phenomena of biology, and its history, are more far more complex – and confusing – than traditional philosophers imagined. Thus, a central theme of the essays that follow is that philosophy of biology must be learned, taught, and thought about by working intensely with "real biology" and serious history of "real biology."

There is, however, a dialectically counterpoised point. The phenomena that biologists study are extremely complex. Valuable insights about complex systems can be gained by working with what scientific modelers sometimes call "toy models." Because philosophers often work with conceptual models and have developed critical tools for this purpose, their training helps them raise well-thought-out questions about the use of models in biology and about how well those models bear on biological knowledge claims. In short, philosophers who are well informed about biology can hope to work closely with biologists in dealing with the various complexities that infect biological work. Philosophical models and the tools of philosophical criticism can help in "locating" problems within larger intellectual contexts, detecting hidden presuppositions and categorizing the advantages or disadvantages of various instruments or model organisms in pursuing certain aims or particular problems. Philosophers can help dissect problems so that they can be rethought in new terms, and they can analyze various ways in which complex units might act as causally integrated wholes. In short, although history and philosophy of biology cannot be done without close contact with biology, historians and philosophers who pay close attention to biology can (at least sometimes) shed light not only on the history and development of biology but also on useful ways of coping with biological problems.

This book is organized in four parts and brings together eleven interrelated essays, written during the last two decades. As such, it does not put forward a neatly unified point of view for it reflects some of the differences of viewpoint among biological disciplines and some of the major developments in biology, which has undergone enormous changes in the last two decades. Nonetheless, the volume is surprisingly unified. It builds on my enduring interest in three related topics: the historical development of work in evolution, genetics, and development; the epistemological issues raised by the interactions among these (and other) disciplines; and the difficulty in achieving conceptual unification of the accounts of the phenomena studied in these biological domains and in biologists' accounts of organisms more generally. It builds to a climax in the concluding chapter, which examines aspects of the ongoing reconception of the ways in which animals are put together thanks to the new integration of development, evolution, and genetics now under way in evolutionary developmental biology.

Part I contains two chapters about general methodological issues: the use of "model organisms" (a term of art of the late twentieth century) in biology and the methodological importance (and difficulty) of "interdisciplinary unification" within biology. Both chapters were written for symposia and address the papers presented at those symposia, but their central arguments can be easily understood without examining the papers in question. These two chapters help set up one of the most fascinating dialectics in biology: the dialectic between the particularity of findings and contingency of organismal traits (which are both reasons for worrying about the feasibility of unification in biology) and the importance of the search for general unification within biology. For those with philosophical interests, one of the main thrusts of this book is that whatever we make of the degree of unification achieved in biology, it is not achieved by establishing a set of general biological laws under which explanations can be subsumed. Organisms embed too much history within them (not only their evolutionary histories but also their individual ontogenies!) for them to fall under laws modeled on those of physics. This illustrates one of the continuing themes of the book, introduced in Part I.

The other three parts of the book focus on evolution, genetics, and development, paying particular attention to the interactions among these disciplines. Because each section has its own introduction, it is not necessary to elaborate upon them here.

Most of the chapters were written for specific occasions. The thematic unity of the book rests, in part, on resistance to excessive reductionism and on a commitment to what might be called "contextualism" – that is, to the importance of understanding the multiple contexts within which knowledge claims, including those of biology, are put forward and the bearing of those contexts on the interpretation of the claims put forward by biologists. Accordingly, most chapters pay close attention to aspects of the historical settings of the biological problems they examine. I seek to help the reader understand how the available biological knowledge and instruments of research yielded challenges to the then-prevalent conceptual tools and theories of the biologists and how, if at all, such difficulties were – or might have been – resolved in context. By returning repeatedly to the interactions among biological disciplines and to the available investigative tools, I reinforce the importance of issues raised by divergences among disciplines. I also locate some reasons for the failure of reductionist programs to provide satisfactory explanations of biological findings by restricting attention to a single level or biological discipline. Throughout the book, I maintain that biology is built on cross-disciplinary interactions that cover micro- to macro evolution and molecules to whole organisms – not just because of the organization of the disciplines themselves but also because organisms, which are thoroughly historical entities, are built via interactions across ontological levels, with both upward and downward causation.

It is in good part thanks to the influence of Marjorie Grene that I have come to recognize the multilayered and multileveled complexity of biological phenomena and the sciences that deal with them and the importance of placing problems – and scientific work – in a proper historical context. If these underlying attitudes are correct, we will not understand the ongoing transformations of biology until we appreciate both *what issues were at stake at particular junctures* and *how difficult it is to match up the concepts employed by scientists working in different disciplines or in different temporal and social settings*.[1] Thus, among the commitments that I hope the reader will take away from this book are the following:

- To understand scientific work properly, one must locate it in its proper intellectual, scientific, and social contexts.
- No single strategy can adequately describe – or prescribe – sound scientific method.
- Reductionism (itself multifaceted and variable in content and style) is one of the most productive organizing (or heuristic) principles for research in biology.
- But, the heuristic power of reductionism often leads scientists to miss important features of the context and organization of biological entities.
- Accordingly, neither the ontology of living beings nor the epistemological difficulties raised by the scientific study of organisms are revealed by focusing exclusively on reductionist strategies in biology or the progress achieved by use of those strategies.
- The clash of disciplinary insights – which include commitments to particular instruments and methods, not just commitments to theoretical positions – is often a key factor in major advances in biology.
- For this reason, cooperative and competitive investigations across disciplines are often essential for the solution of biological problems.
- Correspondingly, understanding of investigations that cross disciplinary boundaries is crucial for understanding how many biological problems are solved.
- The tools of cooperative investigation include the more-or-less philosophical tools involved in reconciling conceptual conflicts – reconciliation accomplished in part by conceptual analysis, in part by empirical and experimental research.

The chapters of this book, written for diverse audiences, highlight the importance of diverse perspectives and bring to bear insights drawn from different sources. The introductions to the four sections help set the chapters into context and provide some guidance about background presuppositions. Accordingly, gentle reader, I hope that you will forgive occasional small overlaps between the chapters and find them instructive rather than tedious.

[1] For some of my more philosophical papers elaborating on these matters, see Burian 1992, 2000, 2001, 2002; Burian and Trout 1995.

And I hope that you will find and correct the errors and oversights that, no doubt, abound in this book.

## REFERENCES

Burian, R. M. 1992. "How the choice of experimental organism matters: Biological practices and discipline boundaries." *Synthese* 92: 151–66.

Burian, R. M. 2000. "On the internal dynamics of Mendelian genetics." *Comptes rendus de l'Académie des Sciences, Paris. Série III, Sciences de la Vie/Life Sciences* 323, no. 12: 1127–37.

Burian, R. M. 2001. "The dilemma of case studies resolved: The virtues of using case studies in the history and philosophy of science." *Perspectives on Science* 9: 383–404.

Burian, R. M. 2002. "'Historical realism,' 'contextual objectivity,' and changing concepts of the gene." In *The Philosophy of Marjorie Grene*, eds. L. E. Hahn and R. E. Auxier. Peru, IL: Open Court Library of Living Philosophers, 339–60.

Burian, R. M., and J. D. Trout. 1995. "Ontological progress in science." *The Canadian Journal of Philosophy* 25: 177–202.



# I

# Methodological Issues

The two chapters in this section, both published in 1993, illustrate the importance of the meetings of the International Society for History, Philosophy, and Social Studies of Biology (ISHPSSB) as a vehicle for interdisciplinary communication. Chapter 2 was first presented in 1991 at an ISHPSSB symposium, which Muriel Lederman and I co-organized, on "the right organism for the job." The papers from the symposium were published as a special section of the *Journal of the History of Biology* (Lederman and Burian 1993). The symposium was inspired in part by an earlier symposium organized by Adele Clarke and Joan Fujimura, held at the 1989 ISHPSSB meeting and later expanded into a book titled *The Right Tools for the Job* (Clarke and Fujimura 1992). Chapter 3 was originally presented at another 1989 ISHPSSB symposium on the importance and difficulty of integration and unification in the biological sciences. The methodological and epistemological issues at the center of all three symposia received wide attention.[1]

The symposium on model organisms focused on a distinctively biological topic in scientific methodology.[2] Experimental work often requires development of an experimental system (Rheinberger 1995, 1997, 1998). In biology, it often depends as well on the domestication and development of an organism especially suited to the problems at hand. For example, when biologists

[1] The electronic version of the Science Citation Index (including the Humanities and Social Science indexes) contains about three hundred citations to the published versions of the three symposia mentioned in this paragraph, including citations to the individual papers as well as the symposia. About two-thirds of the citations are to *The Right Tools for the Job* or its individual chapters.

[2] The literature on model organisms, already large before our symposium, has grown enormously in the last decade, partly because granting agencies have dedicated considerable funds to a small group of specific model organisms. I list here a sampling of recent references of interest to readers of this book: Ankeny 1997; Bolker 1995; Bolker and Raff 1997; Bolker (moderator) et al. 1998; Buntin 1997; Burian 1996; Davis 2000; Dooley and Zon 2000; Gest 1995; Kohler 1994; Logan 1999, 2001; Schneider 2000; Strange 2000; Wayne and Staves 1996.

were limited mainly to breeding experiments and conventional forms of microscopy in studying genes, the relationship of genes to chromosomes could be studied in detail only in organisms with a small number of chromosomes that were easily visualized in standard microscopes and relatively easy to distinguish from one another. In contrast, entirely different requirements had to be met by organisms especially suited to the study of respiratory physiology or to the study of cell lineages (i.e., the developmental history of lines of cells derived from a particular embryonic cell). Thus, starting in the nineteenth century, certain organisms came to be used as models in certain studies. Only later was the use of such models amplified in a way that led to the creation of "model organisms" intended to serve a series of rather general purposes.

Arguably, organisms that can be domesticated in the laboratory represent a biased sample of organisms, with distinctive properties that contribute to rapid turnover, relatively simple ontogenies, and easy standardization – properties that are not characteristic of organisms in general (Bolker 1995; Bolker and Raff 1996; Bolker (moderator) et al. 1998). Thus, on the one hand, to gain access to particular properties of organisms or processes of interest within and between organisms, it is often necessary in practice to use "exceptional" organisms whose exceptional traits make it feasible to study the properties or processes in question. On the other hand, the exceptional character of such model organisms risks fostering systematic bias and puts the generality of the results thus obtained in doubt. The papers in the symposium dealt with these issues and addressed, specifically, historical and methodological issues associated with the use as models of bacterial viruses (Summers 1993), chlamydomonas (a one-celled green alga used, among other things, in the study of photosynthesis) (Zallen 1993), frogs (Holmes 1993), *Drosophila* (Kohler 1993), plant viruses (Lederman and Tolin 1993), and rats (Clause 1993). Chapter 2 is both a general summary of issues raised in the symposium and my attempt to frame some of the key methodological issues raised in those papers.

Chapter 3 was published as part of a symposium on integration in biology (Ruse and Burian 1993). It begins in reaction to van der Steen's criticism of the ideal of unification of the biological sciences and of misapplications of that ideal (van der Steen 1993). The chapter then expands on some of the issues briefly touched on in Chapter 2 by use of a case study of work in the early 1940s leading up to the one-gene – one-enzyme hypothesis. In particular, I explore the importance of the contingencies that arise from the availability of particular tools and techniques and of bringing the experimental and conceptual tools of different disciplines to bear on common or overlapping problems. In the process, I argue that the ideals of resolving conflicting claims and, sometimes, of unifying the conflicting conceptual apparatus of different disciplines serve as important second-order ideals for the development of (biological) sciences. At the same time, I argue that these are weak ideals, for they cannot give specific guidance about the pathway toward resolution or unification. Thus, the ideal does not provide guidance as to whether one or another discipline provides the appropriate basis for unification. Thus, the ideal of unification can be dissociated from the drive toward reduction – even sophisticated weak versions of reduction such as Schaffner's General Reduction–Replacement Model, presented at the same symposium (Schaffner 1993a) and in much greater detail in Schaffner (1993b).

The pairing of these two chapters provides a basis for key arguments about the centrality of "local knowledge"; for example, the partial, open-ended knowledge localized in particular disciplines or based on particular techniques. The contingency of evolution and the openness of the knowledge situation of ongoing scientific investigations both contribute to this conclusion. These themes return in subsequent chapters of the book.

## REFERENCES

Ankeny, R. A. 1997. "The conqueror worm: An historical and philosophical examination of the use of the nematode *Caenorhabditis elegans* as a model organism." Ph.D. dissertation, University of Pittsburgh.

Bolker, J. A. 1995. "Model systems in developmental biology." *BioEssays* 17: 451–5.

Bolker, J. A., and R. A. Raff. 1996. "Developmental genetics and traditional homology." *BioEssays* 18: 489–94.

Bolker, J. A. (moderator), and R. A. Raff. 1997. "Beyond worms, flies, and mice: It's time to widen the scope of developmental biology." *Journal of NIH Research* 9: 35–9.

Bolker, J. A., R. M. Burian, A. Chipman, G. Gibson, M. Martindale, A. R. Morrison, R. A. Raff, A. Sutherland, and R. Wayne. 1998. *Model systems: A debate.* In *HMS Beagle: The BioMedNet Magazine,* Issue 24 (Jan. 30, 1998). Available from http://news.bmn.com/hmsbeagle/24/cutedge/overview.htm.

Buntin, J. D. 1997. "Comparative endocrinology: Avian models." *Endocrine News,* December: 11–12.

Burian, R. M. 1996. "Some epistemological reflections on *Polistes* as a model organism." In *Natural History and Evolution of an Animal Society: The Paper Wasp Case*, eds. S. Turillazzi and M. J. West-Eberhard. Oxford: Oxford University Press, 318–37.

Clarke, A. E., and J. H. Fujimura, eds. 1992. *The Right Tools for the Job: At Work in 20th-Century Life Sciences.* Princeton: Princeton University Press.

Clause, B. T. 1993. "The Wistar rat as a right choice: Establishing mammalian standards and the ideal of a standardized mammal." *Journal of the History of Biology* 26: 329–49.

Davis, R. H. 2000. *Neurospora: Contributions of a Model.* New York and Oxford: Oxford University Press.

Dooley, K., and L. I. Zon. 2000. "Zebrafish: A model system for the study of human disease." *Current Opinion in Genetics and Development* 10: 252–6.

Gest, H. 1995. "*Arabadopsis* to zebrafish: A commentary on 'Rosetta Stone' model systems in the biological sciences." *Perspectives in Biology and Medicine* 39: 77–85.

Holmes, F. L. 1993. "The old martyr of science: The frog in experimental physiology." *Journal of the History of Biology* 26: 311–28.

Kohler, R. E. 1993. "*Drosophila*: A life in the laboratory." *Journal of the History of Biology* 26: 281–310.

Kohler, R. E. 1994. *Lords of the Fly*: Drosophila *Genetics and the Experimental Life*. Chicago: University of Chicago Press.

Lederman, M., and R. M. Burian, eds. 1993. "Symposium on the right organism for the job." *Journal of the History of Biology* 26, 2: 235–367.

Lederman, M., and S. A. Tolin. 1993. "OVATOOMB: Other viruses and the origins of molecular biology." *Journal of the History of Biology* 26: 239–54.

Logan, C. A. 1999. "The altered rationale for the choice of a standard animal in experimental psychology: Henry H. Donaldson, Adolf Meyer, and 'the' albino rat." *History of Psychology* 2: 3–24.

Logan, C. A. 2002. "Before there were standards: The role of test animals in the production of scientific generality in physiology." *Journal of the History of Biology* 35: 329–63.

Rheinberger, H.-J. 1995. "Comparing experimental systems: Protein synthesis in microbes and in animal tissue at Cambridge (Ernest F. Gale) and at the Massachusetts General Hospital (Paul C. Zamecnik), 1945–1960." *Journal of the History of Biology* 28: 387–416.

Rheinberger, H.-J. 1997. *Towards a History of Epistemic Things: Synthesizing Proteins in the Test Tube*. Stanford, CA: Stanford University Press.

Rheinberger, H.-J. 1998. "Experimental systems, objects of investigation, and spaces of representation." In *Experimental Essays: Versuch zum Experiment*. Eds. M. Heidelberger and F. Steinle. Baden-Baden: Nomos.

Ruse, M., and R. M. Burian. 1993. "Integration in biology, a special issue." *Biology and Philosophy* 8: 3.

Schaffner, K. F. 1993a. "Theory structure, reduction, and disciplinary integration in biology." *Biology and Philosophy* 8: 319–47.

Schaffner, K. F. 1993b. *Discovery and Explanation in Biology and Medicine*. Chicago: University of Chicago Press.

Schneider, D. 2000. "Using *Drosophila* as a model insect." *Nature Reviews Genetics* 1: 218–26.

Strange, K. 2000. "Model organisms: Comparative physiology or just physiology?" *American Journal of Physiology. Cell Physiology* 279: C2050–1.

Summers, W. C. 1993. "How bacteriophage came to be used by the phage group." *Journal of the History of Biology* 26: 255–67.

van der Steen, W. J. 1993. "Towards disciplinary disintegration in biology." *Biology and Philosophy* 8, 3: 259–76.

Wayne, R., and M. P. Staves. 1996. "The August Krogh principle applies to plants." *BioScience* 46: 365–9.

Zallen, D. T. 1993. "The 'light' organism for the job: Green algae and photosynthesis research." *Journal of the History of Biology* 26: 269–79.

# 2

# How the Choice of Experimental Organism Matters

## Epistemological Reflections on an Aspect of Biological Practice[1]

> Unless we recognize our innate biases in animal model choice, we limit our potential as experimenters. Two biases seem common from my observations. First is the anthropomorphism that we all seem to get from the monkeys in zoos and circuses, coming as it does long before we aspire to be scientists. Second is for the animal or animals with which we worked during our early days in our fields. Both of these are easy to understand and forgivable. What has neither of these saving attributes is our unwillingness to consider the entire biologic kingdom as a source of possible models of one or another human functions, normal or diseased.[2]

The value of an organism[3] as an experimental tool, or in field studies, depends not only on various features of the organism[4] but also on the problems to be addressed and the available experimental and field techniques. Indeed,

[1] Published in *Journal of the History of Biology* 26 (2) (Summer 1993): 351–67. © 1993 Kluwer Academic Publishers. Reprinted with kind permission of Kluwer Academic Publishers. This was the concluding paper of a special section in the same issue of the journal on "The right organism for the job," eds. M. Lederman and R. M. Burian, 233–367. The section contains papers by Richard M. Burian, Bonnie Tocher Clause, Frederic L. Holmes, Robert E. Kohler, Muriel Lederman, and Sue Tolin, William C. Summers, and Doris T. Zallen. I am grateful to Marjorie Grene, Muriel Lederman, Anne McNabb, Doris Zallen, and a *JHB* referee for their helpful critical readings of a late draft of this chapter.

[2] Prichard (1976, p. 172); used as a chapter epigraph at p. 24 of Committee on Models for Biomedical Research 1985.

[3] I take "organism" in a broad sense here, including such artificial "organisms" as somatic cell lineages in tissue culture. Such biological "material" plays much the same role as an organism for the purposes of this chapter.

[4] This seemingly essentialist *façon de parler* ("the organism") is for convenience only. One of the most important features of "an" organism (i.e., conspecifics, or organisms belonging to a particular strain, perhaps an especially prepared laboratory strain) may be the variability "it" exhibits. Nor is variability always undesirable. Consider, for example, genetic studies of the norm of variation (i.e., the range of phenotypes produced by a given genotype under various circumstances). As this illustration shows, the use or construction of a specially prepared laboratory strain aims

even when some organism is "the" right one for a theoretical job, its rightness is temporary and more or less local or regional.[5] It depends not only on the job but also on the techniques employed and the social or institutional support system for doing that job.

Most biologists realize that the choice of organism can greatly affect the outcome of well-defined experiments and can thus have a major impact on the valuation of biological theories. Anyone who does not appreciate the point need only think of well-known cases in which the choice of organism led investigators down a garden path. Consider three familiar instances: First, recall the difficulties Gregor Mendel faced when he tried to satisfy Carl von Nägeli that his "law for *Pisum*"[6] was generally valid by attempting to apply it to *Hieracium*. Second, consider the strong experimental support for Hugo de Vries's mistaken but not in the least misguided mutationism obtained from the study of *Oenothera* in the first decade of the last century.[7] Third, to cite an animal example, remember the behavior of the chromosomes of the parasitic nematode *Ascaris*. Theodor Boveri discovered "chromosome diminution" in *Parascaris aequorum* (formerly *Ascaris megalocephala*): The chromosomes remain intact in germ-line cells but shatter and are parceled out differently in different somatic-cell lineages. This discovery, made fairly early in the attempt to connect chromosomal behavior to heredity, lent powerful, albeit temporary, support to August Weismann's theory of inheritance.[8]

There must be numerous cases like those involving *Hieracium, Oenothera*, and *Ascaris*, in which an unlucky choice of organism led investigators astray. Such cases are not widely studied because of scientists' and historians'

to limit variation *primarily in those respects that need to be controlled for the purposes of the experiment.*

[5] This point follows easily from Zallen's correct conclusion that a wise choice of organism depends on "the best match of the properties of the organism with the experimental equipment being used" (Zallen 1993, p. 278). Changing techniques, changing background knowledge, the domestication of an alternative organism – these are but some of the changes in the context of knowledge and practice that can alter the "rightness" of an organism for a particular job.

[6] Translations of the original documents into English are collected in Stern and Sherwood (1966). The "misleading" behavior of *Hieracium* is often mentioned although not analyzed in detail in the secondary literature. The failure of *Hieracium* to behave like *Pisum* surely reduced the likelihood that Mendel's work would receive immediate wide attention.

[7] Sturtevant (1971), cited by Zallen (1993), gives a brief account of the unlucky character of de Vries's choice of *Oenothera*. It took extremely painstaking cytological work to learn that plants in this genus have ring chromosomes containing balanced lethal mutations and other aberrations that interfered with "normal" Mendelian behavior. For more technical treatments of these problems, see, e.g., Emerson and Sturtevant (1931) and Cleland (1936).

[8] The phenomenon is briefly described in Gilbert (1988b, pp. 266–7). Cf. Boveri (1887, 1902, 1910).

understandable tendency to dwell on success stories.[9] The papers in the *Journal of the History of Biology* (Lederman and Burian 1993) exhibit this tendency – that is, they focus on influential choices that did not lead into blind alleys. To understand what goes into such successful choices, one should study some contrasting cases – that is, cases in which the choice, alteration, or construction (see Kohler 1993) of organisms to facilitate experimental work failed to have a major impact on the directions in which biological knowledge developed. This important but difficult task is, alas, beyond the grasp of this chapter.

Instead, I explore some generalizations about successful choices and raise some related epistemological questions about the consequences of choice of organism. Because alternative choices of experimental organism can lead in significantly different directions, the epistemological questions are obviously important. They have the potential to significantly alter the evaluation of biological knowledge claims by revealing some of the limitations and qualifications that actual choices place on that knowledge. Our understanding of these limitations will almost certainly be improved when we can compare some significant "failures" in choosing organisms with the better-known "successes."

## FRAMEWORK

This chapter is organized around the following three claims. I draw extensively on the case studies published in the symposium in the *Journal of the History of Biology*, as well as some others presented briefly herein to motivate and support these contentions.

1. Various peculiarities of an organism, some known and some unknown at the beginning of an investigation, significantly affect its suitability for that investigation. It may require a considerable amount of work over a long time to ascertain which of the organism's characters aid and which impede a particular investigation (see Sections 1A and 1B). Because evolution is a branching process with many irregular steps, the extent to which crucial characters occur in other organisms is a question of "evolutionary contingency." Thus, even though on the available evidence some choices are wise and others unwise, scientists cannot be sure in advance that the organisms they have chosen are suitable as a means of investigating their questions, especially if those questions are general ones. This point is

[9] A valuable exception is Mitman and Fausto-Sterling (1992).

related to, but distinct from, the point (briefly discussed in Section 1C) that it takes significant work to "standardize" or "domesticate" organisms and protocols so that results in different laboratories or field studies can be reliably compared. Reliable results about chromosome diminution in *Ascaris* and the distribution of phenotypes in $F_2$ progeny of intraspecific crosses of *Hieracium* or *Oenothera* do not remove the rare features of the physiology of inheritance in each organism.

2. Various attributes of organisms (including mating systems, ecological roles, and infections carried) can transform the investigators' job, forcing them to revise the presuppositions with which they began their investigation and to turn toward domains and findings rather distant from those for which the organism was originally to be employed.
3. The epistemological evaluation of the support for (general) theoretical hypotheses in biology is thoroughly comparative. The use of fieldwork or experimentation with an organism or group of organisms to support theoretical hypotheses requires detailed and substantive knowledge of the special features of the organism(s) and the experimental techniques in question. But, such knowledge is not sufficient to justify theoretical hypotheses because what is known about the experimental organism(s) must also be set into a wider context of knowledge about other organisms and the relevant evolutionary and phylogenetic relationships. Thus, the evaluation of theoretical knowledge in biology is deeply dependent on a broad base of knowledge about the particularities of different organisms and their alternative biochemical mechanisms, life cycles, means of survival, strategies of reproduction, and so on. This is an important consequence of the contingency of evolution and of the fact that organisms and lineages, unlike the fundamental particles of physics, always differ from each other in significant ways.

## 1. THE SUITABILITY OF ORGANISMS

### *A. Traits and Suitability*

The epigraph for this chapter suggests that biologists can draw model organisms from the entire biologic kingdom. Perhaps so – but one should always ask whether (1) the model is faithful in the relevant ways to that for which it is supposed to serve as a model, and (2) it allows a useful analytical approach to the problem at hand. The answers to these questions depend on which aspects of which phenomena are being studied. Some organisms are more adaptable than others, but there are always jobs for which a given organism is unsuited.

Often, alternative organisms satisfy the relevant criteria for doing particular jobs at least as well as the one that is chosen. However, the problems involved in tooling up with a new organism, the advantages of working with one already familiar or used on a large scale, the importance of reliable data on the behavior of the organism and of the availability of reliably characterized strains suited to particular purposes – these considerations bias the evaluation of "suitability" away from an abstract deliberation based merely on the criteria ideally set by the questions to be investigated and the known properties of the relevant organisms. As Frederic L. Holmes (1993) makes clear, there are cases in which it is probably not possible to answer the question whether frogs were the most suitable organisms for certain investigations or simply the most easily available and familiar among reasonably well-suited organisms. The same is true for rats and *Drosophila*, as Bonnie Clause (1993) and Kohler (1993) point out.

In many important cases, phylogenetic proximity is not an appropriate measure of the suitability of a model for a given process. For example, as Doris Zallen (1993) shows, the choice of a suitable organism for studying photosynthesis turned – in the first instance – on solving such physical problems as preventing obstructing layers from interfering with access of light or with diffusion of the gases produced in photosynthesis. Or, to cite a new example, many questions about mammalian ontogeny are difficult to study because of interactions between mother and fetus across the placenta and because of the inaccessibility of the fetus. In at least some instances, birds are better "model organisms" than mammals for studying certain medically relevant aspects of early mammalian or human developmental processes. Bird embryos, after all, experience no direct maternal input once the egg is complete and, therefore, no subsequent maternal influence on biochemical processes in the embryo. Biochemically, as it happens, some endocrinological processes in development are virtually identical in certain birds and mammals. And bird embryos can be studied without the disruptive procedures required for studying mammalian embryos *in utero*.[10] Yet, investigators' familiarity with the ubiquitous rat and discomfort with the relative phylogenetic distance from humans of birds as compared to rats have helped make it difficult for avian models to be taken seriously.

[10] I should reveal my bias. This claim rests partly on my wife's use of avian models to study the role of thyroid hormones in ontogeny. She argues in detail for the preferability of avian models to rats (or sheep, whose ontogeny is in relevant respects closer to humans' than that of rats) as tools for understanding the ways in which thyroid hormones regulate certain aspects of early development, including some that are medically relevant to humans. See, e.g., McNabb (1989); McNabb and King (1993).

On the other hand, even in biochemistry and molecular biology there are cases in which phylogeny reflects, at least in part, critical differences that have arisen in the course of evolution. Thus, to take an extreme example, the structural differences between the genetic material of procaryotes and of eucaryotes are sufficiently great that procaryotes make poor models for many issues concerning gene regulation and gene structure in eucaryotes.[11] This fact frustrated the early optimism of molecular geneticists after the development of the operon model of gene regulation (cf. Jacques Monod's oft-cited quip, "What's true for *E. coli* is true for the elephant"); procaryotes simply are not an adequate model for the regulatory apparatus affecting gene expression in eucaryotes.

The differences among the cases to which I alluded in the last three paragraphs illustrate both the range and the specificity of the considerations that entail the evaluation of a particular organism for a particular job. In the face of the incompleteness of our knowledge of even the best-known organisms, the degree to which particular processes in one can serve as a model of or surrogate for particular processes in another is usually open to question. This is part of the dialectic, noted by Krebs and emphasized by Holmes (1993), between working with organisms that offer special advantages and attempting to gain unobscured access to "basic principles" pertaining, ideally, to large classes of organisms or to organisms in general. Given what we know of the opportunism and tinkering that characterize evolutionary change, we must always be aware of the risk that a series of experiments based on a particular organism deals with a special case. It requires *particular* knowledge of a wide range of relevant cases to evaluate this risk.

### *B. The Complexity of Suitability*

Even when experimental work aims to answer a well-defined question with well-defined materials, it is subject to unexpected contingencies. In biology, the features of the organisms under investigation are a source of some of the most important contingencies involved. (Kohler's account of the *Drosophila* work in T. H. Morgan's laboratory is particularly striking in this regard.)[12] Because of these contingencies, when different organisms are used to investigate the same question, they may yield systematically different results; or, one organism may yield clear-cut results whereas another yields no useful results. Thus, some embryologists sought to answer the fundamental question of

[11] A useful exposition of some of the peculiarities of gene structure and organization in eucaryotes may be found in Hunkapiller et al. (1982).

[12] See particularly Kohler (1994).

whether a differentiated somatic cell nucleus from an adult is totipotent (i.e., able to generate all cell types) by transplanting such nuclei into enucleated eggs. When this was done in several species of sea urchin, no result was obtained because the eggs rejected the foreign nuclei; a few years later, precisely parallel experiments succeeded with frogs.[13] Similarly, seemingly small details of experimental technique – or small changes in experimental protocols or reagent concentrations or purity – can make a major difference in the clarity or even the character of experimental results. Among other consequences of this familiar fact, one is important for our purposes: Small differences in technique can significantly alter the value or the evidential importance of work with a particular experimental organism. What is involved is an interaction among (at least) the questions under investigation, the features of the organisms employed, the skill of the investigators, the experimental tools and techniques available at the time, and the vast panoply of social and institutional factors (of which the latter are underemphasized in this chapter).

No abstract formulation can capture the wealth of relevant details that play a role here – or the different sorts of details that must be considered in different cases. Some suggestive examples help make the point. Summers (1993), for example, implicitly asks why Félix d'Hérelle's arguments that phage should be classed as obligate intracellular parasites met such limited acceptance in the mid-1920s and 1930s. One contributing factor among many, suggested in my interviews with Lwoff and Wollman,[14] was the tendency of many bacteriologists and most students of phage from the 1920s through the 1940s to think of bacterial cultures as uniform wholes. One consequence of this predilection was that a conceptually clear key experiment was not performed for a long time after it was technically possible – namely, starting bacterial cultures from single bacterial cells of so-called lysogenic strains (i.e., strains containing latent phage). The point was to determine whether a *single bacterium*, itself uncontaminated by free phage, could initiate a lysogenic culture – a culture with no free phage that nonetheless could produce active phage – and, if so, whether the kinetics of the phenomenon fit better with d'Hérelle's or Jules Bordet's theory (described by Summers).[15]

[13] This project was attempted, e.g., by P. P. Slonimski as a thesis project ca. 1947–8. According to Slonimski, the problem of rejection of nuclei by sea urchin eggs was not solved until the 1980s. As is familiar, a similar project was carried out successfully by Briggs and King in frogs in the early 1950s (Briggs and King 1952). I am grateful to Dr. Slonimski for describing his project in an interview and for his encouragement of my research.

[14] These interviews were held in 1988 and 1984, respectively.

[15] Lwoff performed an important series of such experiments starting in 1949, leading to the definition of conditions in which phage production could be brought about ("induced") in some

Kohler argues that the scale on which an experiment or a series of experiments is carried out can sometimes be crucial, as it was in the search for mutations in *Drosophila* stocks in Morgan's laboratory. When this is so, secondary factors or secondary uses of the organism can play an important role in determining whether it is employed long enough or on a large enough scale to do the job (here, of finding mutations – but note that "the" job may not be one the investigator had in mind). Beyond the obvious issues of cost, convenience, length of time required for the protocol, and so forth, the organisms' other uses are important. As Kohler shows, the usefulness of *Drosophila* for studies on experimental evolution, for teaching, as a screen for mutations, and so on played an important role in bringing it to the point where the mutation studies overtook the rest. A similar point, though vastly different in detail, derives from Clause's account of the multiple uses of the Wistar rat. Again, the economic importance of organisms (see Lederman and Tolin 1993) or their importance as disease vectors (see Summers 1993) can lead biologists to do a sufficient amount of work with them that they learn of their special virtues for purposes unrelated to those for which they were originally employed.

### C. Conservative versus Risky Strategies in Choosing Organisms

Holmes, Clause, Kohler, and Zallen all emphasize the immense amount of work required to construct and obtain reliably reproducible results from an experimental organism. (Fieldwork also may similarly require heavy investment to establish identification procedures and protocols that allow different investigators to generate reliably reproducible results.)[16] A major part of the effort may go into fine-tuning and standardizing the organism (see Clause 1993; Kohler 1993), matching the strain or organism to the experimental protocols or revising those protocols so that the experiments yield clear results or results that bear directly on the question at hand. Not only must the strains be defined, but the nature of the variation among organisms must also be understood and counteracted where it interferes with the experimental protocol (which it does not always do!). Furthermore, both protocols and organisms must be adjusted

strains of lysogenic bacteria. The first report is Lwoff and Gutmann (1949); the first full account of the series of experiments is Lwoff and Gutmann (1950).

[16] A moderately familiar example helps make the point. In the 1930s, the difficulties in determining the population structure of *Drosophila pseudoobscura*, which seemed to have intersterile races, were ultimately resolved by the identification of a sibling species, *Drosophila persimilis*. The problem of whether the "races" in question were good species was so great – and so theoretically consequential – that the issue played a role in the estrangement between Theodosius Dobzhansky and Alfred Sturtevant. The complex issues involved are well described in Provine (1981).

to one another in the service of the aim of the experiment. Muller's efforts to construct ever more refined laboratory strains of *Drosophila* in the period after he received the Nobel Prize provide an extreme example (see, for examples, Muller 1952, 1953a, and 1953b). During this period, a considerable part of the effort in Muller's laboratory was devoted to constructing esoteric special-purpose strains of *Drosophila* to allow the experimental resolution of extremely specific questions regarding chromosomal mechanics, gene locations, modifier genes, gene regulation, and so on.[17]

Given the heavy investment involved in constructing or gaining control of a laboratory organism[18] (see Clause 1993; Kohler 1993), it is typically quite costly to change organisms, particularly in those instances in which a new organism must be domesticated or adapted to a new purpose. This is not merely a matter of familiarity and comfort; the difficulties encountered are often a function of significant differences in physiology, biochemistry, morphology, genetic content, controls of gene regulation, and so on *ad indefinitum*. Furthermore, when a "large body of experimental information has accumulated" about a particular organism (Zallen 1993, p. 278), that information itself constitutes a resource enabling experimenters to develop protocols and to interpret experimental results in terms of standardized knowledge of the organism. Because, in the end, the behavior of a new organism might prove difficult to interpret and the organism might prove unsuitable for the task at hand, the resistance of some experimental biologists to changing organisms is understandable. In general, it is both easier and safer to stay with a "familiar" organism and exploit it to the full rather than to start working with a less familiar or less fully developed alternative – even when that alternative is potentially well suited to a particular task. For these reasons, biologists often hesitate for an extended period before deciding which new organism, if any, in which to invest. After all, working with a new organism constitutes a major investment, one that usually requires some years before it becomes clear whether it will pay off as promised or lead into a blind alley.

[17] I examined the contrast between this work and that of Milislav Demerec during the same period in a talk, "Model Organisms and Research Strategies in Mendelian Genetics," available on request, delivered at the 17th International Congress of the History of Science, Berkeley, California, 1985. The central point of the contrast concerned Demerec's readiness to switch organisms in hope of finding the "right" organism for pursuing particular problems versus Muller's attempt to adapt "his" organism to the question at hand.

[18] Including such "organisms" as somatic cell lineages. Zallen and Burian (1992, pp. 1–7, esp. pp. 3–5, and references therein) provides a minimal sense of the immense investment required to gain control of such systems and the length of time it takes to learn whether there will be a significant payoff.

To some extent, questions of style and personality enter into the decision whether to stay with an organism or switch to another in pursuing a particular problem. Some investigators are so attached to a particular organism or so confident that they can adapt it to almost any task that, like Muller, they do not seriously consider switching to alternative organisms. Others, motivated by a particular problem, are happy to switch to a more promising organism even at the cost of a few years' labor.[19] Still others, perhaps more adventuresome or more easily bored, fear going stale or missing something because of their commitment to a particular organism or problem. Such individuals are more likely to make a major switch of problems and/or organism every decade or so.[20] It is worth observing that there are almost always serious candidate organisms available, organisms that meet some but not all of the desiderata relevant to the task at hand.[21] Because this is so, and because blind alleys are so common, it is desirable that there be considerable variance among investigators in the degree of inertia (or conservatism) and, conversely, in the willingness to take a chance on tooling up for a different attack on a problem – or for an attack on a new problem.

## 2. TRANSFORMING THE JOB

The symposium articles discussed in this chapter, especially Kohler's, suggest that the interaction between the protocols employed by an investigator and the features of an experimental organism can transform the job undertaken by the investigator. I briefly deploy two examples here to indicate how both the organism and the available techniques can alter the tasks for which a given organism is employed. As the second example shows, it is sometimes possible to remove limitations pertaining to the use of an organism by a change of technique.

[19] As a referee pointed out, investigators may have different underlying assumptions about whether most organisms or only a few very special ones are suitable for investigating a particular problem. Additionally, some problems simply require working with many organisms or a variety of organisms. This is perhaps more common in fieldwork than in experimental work.

[20] To avoid misunderstanding, I should stress that good experimentalists constantly seek to improve their protocols and the match between organism and protocol. The claim in the text concerns major reorientations in experimental programs. I discussed the matter of style briefly in a predecessor to this chapter (Burian 1992, pp. 160 ff.).

[21] The features that an organism should possess to be suitable for a given job are determined in good part by the problem at issue and by the available techniques. Holmes (1993) and Zallen (1993) provide exemplary lists of desiderata appropriate in different cases and circumstances. See also the next section.

1. In the mid-1930s, Philippe L'Héritier and Georges Teissier performed the first population cage experiments with *Drosophila*. They showed, among other things, that unfavorable mutations (e.g., *bar*) could be maintained in a population in a stable balanced polymorphism despite the deleterious effects of the mutation on the flies that carried it (see, e.g., L'Héritier and Teissier 1933, 1937). Such experiments required a rapid way of counting thousands of flies without disrupting the population or altering its numbers. For this purpose, a useful device was to anesthetize the flies briefly with $CO_2$, spread them on a photographic plate, and count images rather than flies. As it turned out, in some populations, a substantial proportion of the flies was killed by this procedure. Given their plan, the experimenters had to gain control of their counting technique. L'Héritier made a special study of this phenomenon. The sensitivity to $CO_2$ proved not only to be heritable *but also to be maternally inherited* – that is, to be produced by some genetic factor other than a chromosomal gene, most likely contained in the cytoplasm deposited into the egg by the mother. Eventually, the need to understand and gain control of this $CO_2$ sensitivity came to play a central role in L'Héritier's research; after twenty years or so (interrupted by World War II), he established that it was due to a maternally transmitted viroid, since labeled sigma, pandemic in most *Drosophila* populations (L'Héritier 1951, 1970). The behavior of the fly, altered by an infection, had displaced L'Héritier from population genetics to a study of cytoplasmic inheritance.
2. Staying with *Drosophila*, one aspect of the well-known gulf between embryology and genetics from the 1920s through the 1960s[22] concerned the requirements placed on experimental organisms by the research agendas of these two disciplines in light of their respective classical techniques. To study early ontogeny, embryologists needed large, transparent, manipulable eggs that formed a series of visibly distinct cells and easily marked cell lineages. Given these and other favorable conditions, embryologists could follow the pathways by means of which various tissues and organs arose and could study the effects on ontogeny of various manipulations, transplantations, and chemical insults. Although maize meets most of these conditions, virtually no animal then employed in or known to be suited to classical genetic experimentation does. An ideal genetic organism

[22] There is, by now, a large literature on this topic. For examples, see Allen (1986); Clarke (1991); Gilbert (1978, 1988a); Mitman and Fausto-Sterling (1992); Sapp (1987); and numerous writings of Jane Maienschein. See also Chapter 11.

would be easily raised in the laboratory, allow controlled matings to yield multigeneration pedigrees, have a relatively short generation time, be able to tolerate both inbreeding and outbreeding, have a good supply of well-known and easily scored mutations, and, by the 1920s, have a small number of visible and easily distinguished chromosomes. Take *Drosophila*, the classical genetic organism, as an example. It is badly suited to classical embryology: The egg is small, difficult to manipulate, and encased in a tough opaque coating; the first hundred or so nuclei, identical in appearance, form a syncytium (i.e., a single cell with multiple-cell nuclei) before cell walls are formed. That dramatic change proceeds in parallel throughout the embryo over a brief period and is difficult to follow; even then, most of the resultant cells are not well marked. There are further disadvantages as well, but this is enough to make it clear that *Drosophila* is poor material for classical embryology.[23]

In recent years, however, this situation has changed dramatically. A wealth of new molecular techniques (e.g., in situ hybridization and immunofluorescence) allows students of development to mark rare molecules and follow their distribution over short periods during ontogeny. In addition, there are powerful new techniques for altering and inactivating the genes that produce those molecules. Together, these techniques have allowed step-by-step visualization and dissection of the earliest signals that lay down heretofore-invisible boundaries within *Drosophila* eggs and early larvae. Recently, there has been enormous progress in analyzing the controls by means of which these boundaries affect the ontogeny of the organism (see Chapter 12). In effect, the conditions described previously are now being met – it is now possible to visualize occurrences behind the integument of the egg, to manipulate and perturb the egg from its normal state in specific ways, and to mark and follow cell lineages. Joining the new techniques to the immensely powerful preexisting genetic practice and information available for *Drosophila* has made this organism into a preferred organism for the study of development. Thus, in this instance, an organism that used to be wholly unsuited for a major task (i.e., the study of early development) has become particularly suited for that job, bringing with it a major new set of

[23] A significant number of biologists in the 1920s and 1930s sought organisms suited to both embryological and genetic work or to employ a "genetic" organism in embryological studies or vice versa. For studies of one biologist who attempted to bridge this gap using *Drosophila* inter alia, see Burian, Gayon, and Zallen (1988, 357–402; esp. pp. 389–407, 1991). A number of the other studies in the latter volume bear on this topic.

evolving experimental practices.[24] This illustrates how new techniques can bring about remarkable changes in the jobs that can be performed with an organism. It also illustrates the value of an enormous database and of well-developed experimental practices with a given organism. Finally, it shows how great a length of time (here, at least thirty years!) may be involved in adapting an organism and experimental techniques to a task for which the organism was not initially suited. It is no wonder that many experimentalists find it more productive to continue working with and adapting their favorite organism incrementally to new jobs.

## 3. THE BREADTH OF THE BIOLOGICAL KNOWLEDGE CONTEXT

Evolution is a branching process in which each organism (each lineage, each species) has distinct characters, differing in at least some way from the organisms (lineages, species) from which it stemmed. This means that each organism provides an imperfect window on the properties of other organisms. Mutation and epigenetic difference produce divergence within lineages. This happens even with contemporary techniques of cloning, which – for relatively simple organisms – can come extremely close to producing organisms with identical properties.[25] This truism proves to be of great importance. At (virtually?) all levels of the biological world – including the biochemical – the generality of the findings produced by use of a particular organism is open to question. Consider a few illustrative examples of the "depth" of this point. The genetic code is not universal – some ciliated protozoa and mitochondria translate certain codons differently than in the "universal" code. (The differences in translation depend on the population of transfer RNAs in the surrounding medium.) Some organisms (or "organisms" – to wit, viruses) have no metabolism of their own. There are organisms (viruses) whose genetic material is not DNA. There may be organisms whose genetic material includes units composed of proteins ("prions"). A given strand of DNA, even

[24] For those unfamiliar with these developments, a useful book summarizing recent work is Lawrence (1992). The organization of topics in this volume indicates the extent to which classical embryological questions are yielding to the new techniques: the first six chapter headings are "The Mother and the Egg" (including gastrulation and segmentation), "The First Coordinates," "Patterning the Embryo," "Cell Lineage and Cell Allocation," "Positional Information and Polarity," and "Spacing Patterns."

[25] In (most?) mammals, the ontogeny of the immune system guarantees that even identical twins will have genetically distinct somatic cells and immune responses. This illustrates the difficulty of obtaining full reproducibility in biological systems.

when it is transcribed to RNA, may yield different information according to the biochemical context (e.g., because of "overlapping reading frames," because the "readout" of DNA sometimes stops at one point and sometimes another, or because a given premessenger RNA is spliced in alternative ways before being translated).[26]

At every level of biological organization, there are innumerable results like these. This makes epistemological evaluation of experimental work in biology especially difficult. To what extent are the results obtained with an organism (or a group of organisms) general and to what extent can they be reliably extrapolated? This is an especially acute version of the traditional philosophical problem of induction. In light of considerations like those just offered, we know we are dealing with a congeries of contingently different systems whose regularities, even if they trace back to fundamental (e.g., biochemical) laws, depend at least as strongly on the contingent, evolutionarily derived, configurations of the components of the system as they do on those laws. In this respect, the epistemological situation of biology is different from that of any form of mechanics, including quantum mechanics, for *biological knowledge is knowledge of large numbers of particular systems that cannot be identically prepared.* As such, it cannot, in principle, be derived from a body of laws plus initial or boundary conditions. Substantive knowledge of evolutionary history, of alternative biological mechanisms, of phylogenies, and so on, is needed to evaluate the power of a given result.

Like Kauffman (1993) and many others, one can attempt a statistical mechanics of complex systems of a general class within which organisms fall (see also Burian and Richardson 1991). *For specific biological knowledge, however, one needs, in addition, knowledge of the historical (i.e., evolutionary) processes that change or can change the properties underlying the statistics.* At every level, the contingency of evolutionary processes must be considered.[27] It follows that, in the end, proper evaluation of the knowledge gained by working with a given organism or group of organisms requires that knowledge to be set into a comparative and evolutionary framework. This is part

[26] Except for the claim about prions, all of these claims are uncontroversial and can be documented adequately in virtually any good current textbook of molecular biology or molecular genetics. See also Chapters 9 and 12.

[27] [Added in 2003;] For a philosophical elaboration on evolutionary contingency, see Beatty (1995), Carrier (1995), and Schaffner (1995). For biological elaboration, see Gould (1986, 1989) and the many related debates over the evolutionary radiations that take place after mass extinctions and over what might happen with a "resetting of the biological clock."

of what I mean by the breadth of the knowledge context in biology. If my account is correct, detailed knowledge of (historical) biological contingencies constrains – and ought to constrain – the evaluation of experimental work in biology and the knowledge claims based on that work.

This stance is reinforced by considerations regarding the interaction between choice of organism and choice of experimental technique or tools. The practices, problems, and epistemological prospects of a discipline can be altered as radically by a change of experimental tools as by a change of organism. Consider the suggestion in the previous section that the epistemological situation of embryology (significantly relabeled "developmental biology") has been improved by the introduction of contemporary molecular techniques. The improved prospects are not unconnected with the fact that molecular techniques have allowed many embryologically intractable organisms such as *Drosophila* to do work in developmental biology. Similar points can be made about population genetics; the transformation of that discipline's treatment of genetic variation when it was invaded by the technique of gel electrophoresis (originally a tool of protein biochemistry) is relatively familiar. The more recent switch from tools for analyzing proteins to tools for analyzing DNA has brought about an equally dramatic transformation. Thanks to the new techniques, many formerly unresolvable questions about variation (e.g., about the extent of neutral variation and the extent to which selection has eliminated variants) have become much more tractable.[28] For our purposes, the moral is straightforward: It is not possible to carry out a reliable epistemological evaluation of work of this type without deep and detailed knowledge of the methodologies employed, the features of the organisms studied, and the match up (or interaction) between the two. It is not a matter of principle but of biology that there is more information about the history and the variation of organisms contained in DNA than in proteins. Such biological knowledge is an inescapable component of the sound evaluation of biological knowledge claims. Epistemological evaluation of biological knowledge claims must employ some sort of bootstrapping technique, for it must rest, in part, on biological knowledge.

[28] Lewontin (1974) provides an epistemologically sophisticated account of the problems that were supposed to be solved by the use of gel electrophoresis, and the then-apparent limitations of the technique for this purpose. Lewontin has recently argued that work at the DNA level can resolve many of the problems that could not be settled by electrophoresis. The principal reason is not limitations of the technique as such (although they are important) but rather the much greater information about variation contained in DNA molecules as compared with protein molecules.

REFERENCES

Allen, G. E. 1986. "T. H. Morgan and the split between embryology and genetics, 1910–1935." In *A History of Embryology*, eds. T. J. Horder, J. A. Witkowski, and C. C. Wylie. Cambridge: Cambridge University Press, 113–46.

Beatty, J. 1995. "The evolutionary contingency thesis." In *Concepts, Theories, and Rationality in the Biological Sciences: The Second Pittsburgh-Konstanz Colloquium in the Philosophy of Science*, eds. G. Wolters and J. Lennox, in collaboration with P. McLaughlin. Konstanz and Pittsburgh: Universitätsverlag Konstanz and University of Pittsburgh Press, 45–81.

Boveri, T. 1887. "Die Bildung der Richtungskörper bei *Ascaris megalocephala* und *Ascaris lumbricoides*." *Jenaische Zeitschrift für Naturwissenschaft* 21: 423–515.

Boveri, T. 1902. "Die Befruchtung und Teilung des Eies von *Ascaris megalocephala*." *Jenaische Zeitschrift für Naturwissenschaft* 22: 685–882.

Boveri, T. 1910. "Über die Teilung centrifugierter Eier von *Ascaris megalocephala*." *Wilhelm Roux Archiv für Entwicklungsmechanik der Organismen* XXX: 101.

Briggs, R. W., and T. J. King. 1952. "Transplantation of living nuclei from blastula cells into enucleated frogs eggs." *Proceedings of the National Academy of Sciences, USA* 38: 455–63.

Burian, R. M. 1992. "How the choice of experimental organism matters: biological practices and discipline boundaries." *Synthese* 92: 151–66.

Burian, R. M., and R. C. Richardson. 1991. "Form and order in evolutionary biology: Stuart Kauffman's transformation of theoretical biology." *PSA 1990* 2: 267–87.

Burian, R. M., J. Gayon, and D. Zallen. 1988. "The singular fate of genetics in the history of French biology, 1900–1940." *Journal of the History of Biology* 21: 357–402.

Burian, R. M., J. Gayon, and D. Zallen. 1991. "Boris Ephrussi and the synthesis of genetics and embryology." In *A Conceptual History of Embryology*, ed. S. Gilbert. New York: Plenum, 207–27.

Carrier, M. 1995. "Evolutionary change and lawlikeness: Beatty on biological generalizations." In *Concepts, Theories, and Rationality in the Biological Sciences: The Second Pittsburgh-Konstanz Colloquium in the Philosophy of Science*, eds. G. Wolters and J. Lennox, in collaboration with P. McLaughlin. Konstanz and Pittsburgh: Universitätsverlag Konstanz and University of Pittsburgh Press, 83–97.

Clarke, A. E. 1991. "Embryology and the rise of American reproductive sciences, circa 1910–1940." In *The Expansion of American Biology*, eds. K. R. Benson, J. Maienschein, and R. Rainger. New Brunswick, NJ: Rutgers University Press, 107–32.

Clause, B. T. 1993. "The Wistar rat as a right choice: Establishing mammalian standards and the ideal of a standardized mammal." *Journal of the History of Biology* 26: 329–49.

Cleland, R. E. 1936. "Some aspects of the cytogenetics of *Oenothera*." *Botanical Review* 2: 316–48.

Committee on Models for Biomedical Research, Board on Basic Biology, Commission on Life Sciences, National Research Council (USA) 1985. *Models for Biomedical Research*. Washington, DC: National Academy Press.

Emerson, S., and A. H. Sturtevant. 1931. "Genetic and cytological studies on *Oenothera*. III. The translocation interpretation." *Zeitschrift für induktive Abstammungs- und Vererbungslehre* 59: 395–419.
Gilbert, S. F. 1978. "The embryological origins of the gene theory." *Journal of the History of Biology* 11: 307–51.
Gilbert, S. F. 1988a. "Cellular politics: Ernest Everett Just, Richard B. Goldschmidt, and the attempt to reconcile embryology and genetics." In *The American Development of Biology*, eds. R. Rainger, K. R. Benson, and J. Maienschein. Philadelphia: University of Pennsylvania Press, 311–46.
Gilbert, S. F. 1988b. *Developmental Biology*, 2nd ed. Sunderland, MA: Sinauer.
Gould, S. J. 1986. "Evolution and the triumph of homology, or why history matters." *American Scientist* 74: 60–9.
Gould, S. J. 1989. *Wonderful Life: The Burgess Shale and the Nature of History*. New York: Norton.
Holmes, F. L. 1993. "The old martyr of science: The frog in experimental physiology." *Journal of the History of Biology* 26: 311–28.
Hunkapiller, T., H. Huang, L. Hood, and J. H. Campbell. 1982. "The impact of modern genetics on evolutionary theory." In *Perspectives on Evolution*, ed. R. Milkman. Sunderland, MA: Sinauer Associates, 164–89.
Kauffman, S. A. 1993. *Origins of Order: Self-Organization and Selection in Evolution*. New York: Oxford University Press.
Kohler, R. E. 1993. "*Drosophila*: A life in the laboratory." *Journal of the History of Biology* 26: 281–310.
Kohler, R. E. 1994. *Lords of the Fly:* Drosophila *Genetics and the Experimental Life*. Chicago: University of Chicago Press.
Lawrence, P. A. 1992. *The Making of a Fly: The Genetics of Animal Design*. Oxford: Blackwell Scientific Publications.
Lederman, M., and R. M. Burian, eds. 1993. "*The right organism for the job*." *Journal of the History of Biology* Vol. 26, 2: 235–367.
Lederman, M., and S. A. Tolin. 1993. "OVATOOMB: Other viruses and the origins of molecular biology." *Journal of the History of Biology* 26: 239–54.
Lewontin, R. C. 1974. *The Genetic Basis of Evolutionary Change*. New York: Columbia University Press.
L'Héritier, P. 1951. "The $CO_2$ sensitivity problem in *Drosophila*." *Cold Spring Harbor Symposia on Quantitative Biology* 15: 99–112.
L'Héritier, P. 1970. "*Drosophila* viruses and their role as evolutionary factors." *Evolutionary Biology* 4: 185–209.
L'Héritier, P., and G. Teissier. 1933. "Une expérience de sélection naturelle: Courbe d'élimination du gène 'bar' dans une population de *Drosophiles* en équilibre." *Comptes rendus de la Société de Biologie* 114: 1049–51.
L'Héritier, P., and G. Teissier. 1937. "L'élimination des formes mutantes dans les populations de *Drosophiles*." *Comptes rendus de l'Académie des Sciences, Paris* 205: 1099–101.
Lwoff, A., and A. Gutmann. 1949. "Les problèmes de la production du bactériophage par les souches lysogènes. La lyse spontanée du *Bacillus megatherium*." *Comptes Rendus de l'Académie des Sciences, Paris* 229: 605–7.

Lwoff, A., and A. Gutmann. 1950. "Recherches sur un Bacillus megatherium lysogène." *Annales de l'Institut Pasteur* 78: 711–39.

McNabb, F. M. A. 1989. "Development and aging of the thyroid in homeotherms." In *Development, Maturation, and Senescence of Neuroendocrine Systems: A Comparative Approach*, eds. M. P. Schreibman and C. G. Scanes. New York: Academic Press, 333–51.

McNabb, F. M. A., and D. B. King. 1993. "Thyroid hormone effects on growth, development, and metabolism." In *The Endocrinology of Growth, Development, and Metabolism in Vertebrates*, eds. M. P. Schreibman, C. G. Scanes, and P. K. T. Pang. San Diego, Academic Press, 393–417.

Mitman, G., and A. Fausto-Sterling. 1992. "Whatever happened to Planaria? C. M. Child and the physiology of inheritance." In *The Right Tools for the Job: At Work in 20th-Century Life Sciences*, eds. A. E. Clarke and J. H. Fujimura. Princeton: Princeton University Press, 172–97.

Muller, H. J. 1952. "Breeding systems for detection of sex-linked lethals in successive generations." *Drosophila Information Service* 26: 113–14.

Muller, H. J. 1953a. "Autosomal mutation studies by means of crisscrossed lethals and balanced male-steriles." *Drosophila Information Service* 27: 104–5.

Muller, H. J. 1953b. "Autosomal nondisjunction associated with the rotund translocation." *Drosophila Information Service* 27: 106–7.

Prichard, R. W. 1976. "Animal models in human medicine." In *Animal Models of Thrombosis and Hemorrhagic Disease*. Washington, DC: U.S. Department of Health, Education, and Welfare, 169–72.

Provine, W. B. 1981. "Origins of the genetics of natural populations series." In *Dobzhansky's Genetics of Natural Populations, I–XLIII*, eds. R. C. Lewontin, J. A. Moore, W. B. Provine, and B. Wallace. New York: Columbia University Press, 1–83.

Sapp, J. 1987. *Beyond the Gene: Cytoplasmic Inheritance and the Struggle for Authority in Genetics*. New York: Oxford University Press.

Schaffner, K. F. 1995. "Comments on Beatty." In *Concepts, Theories, and Rationality in the Biological Sciences: The Second Pittsburgh-Konstanz Colloquium in the Philosophy of Science*, eds. G. Wolters and J. Lennox, in collaboration with P. McLaughlin. Konstanz and Pittsburgh: Universitätsverlag Konstanz and University of Pittsburgh Press, 99–106.

Stern, C., and E. R. Sherwood, eds. 1966. *The Origin of Genetics, a Mendel Source Book*. San Francisco: W. H. Freeman and Co.

Sturtevant, A. H. 1971. "On the choice of material for genetical studies." *Stadler Genetics Symposia* 1: 51–7.

Summers, W. C. 1993. "How bacteriophage came to be used by the phage group." *Journal of the History of Biology* 26: 255–67.

Zallen, D. T. 1993. "The 'light' organism for the job: Green algae and photosynthesis research." *Journal of the History of Biology* 26: 269–79.

Zallen, D. T., and R. M. Burian. 1992. "On the beginnings of somatic cell hybridization: Boris Ephrussi and chromosome transplantation." *Genetics* 132: 1–8.

# 3

# Unification and Coherence as Methodological Objectives in the Biological Sciences[1]

> The biochemist cannot understand what goes on chemically in the organism without considering genes any more than a geneticist can fully appreciate the gene without taking into account what it is and what it does. It is a most unfortunate consequence of human limitations and the inflexible organization of our institutions of higher learning that investigators tend to be forced into laboratories with such labels as "biochemistry" or "genetics." The gene does not recognize the distinction – we should at least minimize it (Beadle 1945b, p. 193).

In this chapter, I respond to Wim van der Steen's (1993) arguments against the supposed current overemphasis on norms of *coherence* and *interdisciplinary integration* in biology. On the normative level, I argue that these are *middle-range norms* that – although they may be misapplied in short-term attempts to solve (temporarily?) intractable problems – play a guiding role in the longer-term treatment of biological problems. This stance is supported by a case study of a partial success story, the development of the one-gene–one-enzyme hypothesis. As that case shows, the goal of coherent interdisciplinary integration not only provides guidance for research but also provides the standard for recognizing failed integrations of the sort that van der Steen criticizes.

[1] Published in *Biology and Philosophy*, 8 (1993): 301–18. © 1993, Kluwer Academic Publishers, as part of a symposium on integration in biology (Ruse and Burian 1993). Reprinted by kind permission of Kluwer Academic Publishers. Work on this paper was supported by a generous grant from the National Endowment for the Humanities. The paper has been improved by criticisms from the other symposiasts and numerous colleagues; by discussion at the International Society for the History, Philosophy, and Social Studies of Biology; and by some improvements in my treatment of the case study suggested by Norman Horowitz. I am grateful to all concerned for their support and assistance.

## INTRODUCTION

This chapter is a defense against the challenge of van der Steen (1993) to the view that the norm of unification of knowledge across (biological) disciplines serves, and *should* serve, as a major vehicle for improving the content of (biological) knowledge. I also deal briefly with the views of Bill Bechtel (1993) and Ken Schaffner (1993a) in my discussion of van der Steen's position.

Van der Steen (1990, 1993; see also van der Steen and Sloep 1988; van der Steen and Thung 1988) argues forcefully against premature and misguided attempts to unify biological disciplines or theories. His argument is directed partly against misuse of methodologies that emphasize coherence between theories or require concordance in the descriptive and theoretical terminology employed by different disciplines. It is also directed generally against the role that such methodologies play in contemporary biology.

There is much of merit in van der Steen's treatment of particular cases. However, I argue that his normative stance – to wit, that the ideal of unification is often inappropriate and damaging even in cases where different disciplines address the same domain of problems or the same range of cases – is fundamentally mistaken. The issue between us turns, in part, on two delicate questions: (1) What should count as the same domain of problems? and (2) What is the proper scale on which to judge and to employ the norms relevant to coherence and unification in science? My position is that the overlap of problem domains must be locally resolved (i.e., by analyzing the specific content of the theories or disciplines at issue and, perhaps, by performing appropriate experiments) and that coherence and unification are norms of the middle range. Thus understood, I argue, those norms are crucial for the development of biological knowledge.

The central issues in the papers presented in the original symposium from which this chapter derives (Bechtel 1993, Schaffner 1993a, van der Steen 1993) concern normative methodology, not the description of actual reasoning patterns or the particular reasons for success and failure in particular cases. Unless I am mistaken, the four of us are generally (though not universally) in accord regarding the analysts of those case studies about which our knowledge overlaps. Yet, although we support our normative claims by reference to just such case studies, we reach divergent conclusions. Perhaps we suffer from sample bias; that is, perhaps our choice of cases may have contaminated our intuitions. Perhaps our disagreements about normative methodology arise because we are asking different questions or because of unexplored normative differences of general philosophical interest. Let the reader beware!

## A FRAMEWORK FOR DISCUSSION

Let us start from simple intuitions. Different schemes of classification and description are employed for different purposes. We are often – but by no means always and never entirely – free to use arbitrary concepts and notions designed to suit the purposes at hand. The value of a particular scheme depends *both* on the purpose(s) at hand and on available knowledge and techniques. Many of the purposes common in biology (e.g., description of a physiological mechanism, establishment of a phylogeny, or application of evidence from different sources to evaluate competing theoretical accounts of a common problem) place specific constraints on the procedures and, ultimately, the concepts that may properly be employed. The constraints discussed in this chapter are long-range but local in ways emphasized by Bechtel (1993);[2] they ought not to be taken too seriously in the short term.

Biological purposes like those just mentioned require that we ultimately (though by no means immediately):

1. achieve coherence among different scientific descriptions of the phenomena of concern and also among those descriptions and any theoretical explanations of those phenomena, or
2. transform the problem of concern so as to remove the obligation to take into account one or more of the competing descriptions or theories.

In the context of van der Steen's concerns about interdisciplinary integration, it is worth pointing out that (1) applies to both intra- and interdisciplinary situations. The ideal of unification requires (at least) that discordant descriptions of the same phenomena, whether or not stemming from different disciplines, be brought into line with one another. Bechtel (1993) provides an interdisciplinary example: Once major respiratory functions had been localized in mitochondria, it was necessary to achieve concordance between structural descriptions of mitochondria and descriptions of the behavior of the relevant respiratory enzymes. The most useful extended case study of the techniques of theory integration in biology known to me is Lindley Darden's book on the interactions between cytology and Mendelian genetics (Darden 1991). Her work reinforces the position I advocate here: Scientists

[2] All first-order constraints are relative to the knowledge context. (For the term "local constraints" in another context, cf. Maynard Smith et al. 1985.) Those who are uncomfortable with this contextual aspect of knowledge claims and methodologies should remember that it is impossible to detach scientists (or any other knowing beings) from the world within which they live, operate, and evaluate their beliefs and theories.

should attempt to make differing theoretical accounts of a domain of phenomena cohere with one another and with well-supported descriptions of those phenomena whether the theories stem from one or more disciplines.

Sometimes an important part of the process of reconciliation involves ascertaining that what seem, *prima facie*, to be different descriptions of the same phenomenon are descriptions of distinct phenomena. One way of transforming a problem is by dividing it into strongly separated problems. For example, the rejection of Haeckel's biogenetic law (Haeckel 1876) facilitated just such a separation between the problem of determining the causes of evolution and the problem of establishing phylogenies. Alternatively, problem transformation may call for a radical reconception of the problem that greatly alters its relation to "neighboring" problems. This happened to the problem of differentiation when it was recognized that determination and differentiation are often independent steps that frequently take place at widely separated stages of the life cycle of an organism or in the history of a cell lineage.

One cannot expect to solve difficult problems quickly; nonetheless, there is such a thing as failure. If, after a suitable period (where suitability depends on available technique, effort expended, and many aspects of the knowledge context), the relevant scientific communities fail to accomplish (1) or (2) they should, *in the end*,

3. confess failure.

I do not think that one can escape the dilemmas implicit in this stance. Failure to reconcile the claims at stake in a dispute concerning a particular subject matter means either that the appearance of a single subject matter is false or that the disputing parties have failed to reach agreement about their common subject. If the dispute falls in the category of disputes about what is the case, these choices are both exclusive and exhaustive.

These considerations lead rather immediately to *qualified* and *contingent* norms and methodological prescriptions. These norms are long-term rather than immediate in a variety of ways. For example, they say nothing about the relative methodological priority of the conflicting protocols or the likelihood of truth of the conflicting theoretical assumptions of different disciplines, nor about the grounds for preferring one terminology or mode of description to another.[3] Again, it is a matter for judgment based on the local

[3] There *are* things to say about such priorities and about how to estimate the likelihood of one theory being nearer to the truth than another. But the dependence of those things on the specific knowledge context ensures the local character of the norms governing sound methodology. In this respect, the more formal and abstract the statement of a norm, the more that norm functions like a meta-level constraint on more immediate norms.

knowledge context whether the available means for resolving the issue have yet been adequately deployed. In this sense, the norms that I describe are weak – and are often legitimately perceived to be weak by bench scientists and students of laboratory life. Nonetheless (except by simply abandoning a seemingly pressing problem, they are difficult to remove; *on an appropriate scale*, precisely because they yield continuing constraints they do and *ought to do* a great deal to shape the direction and outcome of research in the middle term.[4] An analogy demonstrates the power of weak forces to dominate strong ones in appropriate circumstances: The gravitational force is twenty-three orders of magnitude weaker than the next weakest of the known fundamental forces in physics; yet, gravity is the main gestalting agent of the universe on an astronomical scale. Whether or not norms of the middle range have shaping power in science cannot be settled by short-term studies.

## A BRIEF CASE STUDY

To keep this rather abstract discussion attached to real biology – as we must if we are to make headway with the main issues at stake in this chapter – one should work through a number of case studies. Alas, given space limitations, I can deal only with a single case, treated in telegraphic style. That case, however, will help considerably in mobilizing support for the norms of the middle range defended later – and it yields some suggestions about ways in which to meet van der Steen's strictures about biased choice of case studies.

Let me address those strictures briefly. Van der Steen is right, of course, that it is misleading to pay attention primarily to successful cases of interdisciplinary integration. Surely, the vast majority of attempts to achieve integration fail. Nonetheless, we will see that "successful" integration involves many sidetracks and missteps. Careful examination of such cases shows how use of the *ideal* of integration provides important tools for correcting mistaken beliefs and for replacing less informative with more informative practices, protocols, and theories. In short, *in the face of uncertainty about whether an attempt at integration will succeed, one can obtain guidance about how to proceed by examining the means by which mid-course corrections have been made in instances in which successful integrations were achieved.*

[4] By the middle term, I mean to exclude a scale of days, weeks, or months, and also most interactions within a single laboratory, but to include the interactions between laboratories and disciplines on a scale of months, years, and, perhaps in extreme cases, decades.

Enough preliminaries! The case concerns the so-called one-gene–one-enzyme hypothesis. Most readers are familiar with the basic idea, but few know the early formulations of this hypothesis in the classical work of Beadle, Tatum, et al. on *Neurospora*. This work broke new ground by establishing the first extensive correlations between defects in intermediary metabolism and gene mutations. Such correlations were crucial to the development of genetics because (with few partial exceptions) the complexity of genotype–phenotype interactions had hitherto prevented detailed analysis of the physiological steps affected or controlled by the actions of particular genes (see the quotations in Box One).

In fact, as originally formulated, the one-gene–one-enzyme hypothesis was as wrong in its details as were Mendel's laws – for good and interesting reasons. Among these, I focus on two, concerning presuppositions about the nature of genes and the process of enzyme synthesis. There are two orthodoxies in the background, presupposed in many early (rather inarticulate) formulations of the one-gene–one-enzyme hypothesis. It is easy to show that both presuppositions were well supported at the time (i.e., 1941–5), although the first was beginning to be seriously challenged by the end of that period.

Most geneticists believed that proteins had to be the key components of genes; genes were either proteins (perhaps, more specifically, enzymes) or nucleoproteins but, if the latter, the protein components had to provide genes with the specificity required to control the synthesis of different enzymes (see Kay 1992). Nucleic acid was a boring tetranucleotide which, at best, provided a structural framework helping space amino acids during protein synthesis, thus allowing the protein component to assume and retain a stable configuration for it to serve as a template in the formation of enzymes with specific activity.[5]

This doctrine interlocked with a second: namely, that a single precursor could yield many enzymes by altering its conformation in response to a template – sometimes involving the substrate, sometimes involving genes, sometimes both.[6] There were strong experimental arguments against supposing

[5] Cf. Astbury (e.g., 1938), whose advocacy of the role of nucleic acid as a structural frame is cited favorably by Wright (1945). The notion that genes serve as templates does not depend on Astbury's account of the role of nucleic acid, as is clear from the quotations from Gulick, Beadle, and Wright in Box Two.

[6] Monod (1947) articulates this widely shared view. Spiegelman (1950) reviews the extraordinary range of potential mechanisms of enzyme synthesis still open (including template-driven conformational changes). N. Horowitz cautions (personal communication, January 28, 1992) that the template theory was "equally consistent [with] the idea that many different polypeptides were available in the cell, and that the gene selected the one, or ones, that it could interact with and produce an enzyme." He suggests that this option was favored by Beadle and Wright.

BOX ONE

*Prior Limitations on Knowledge of Gene–Enzyme Correlations*

From the standpoint of physiological genetics, the development and functioning of an organism consist essentially of an integrated system of chemical reactions controlled in some manner by genes. It is entirely tenable to suppose that these genes[,] which are themselves a part of the system, control or regulate specific reactions in the system either by acting directly as enzymes or by determining the specificities of enzymes. . . . In investigating the roles of genes, the physiological geneticist usually attempts to determine the physiological and biochemical bases of already known hereditary traits. . . . There are, however, a number of limitations inherent in this approach. Perhaps the most serious of these is that the investigator must in general confine himself to a study of nonlethal heritable characters. Such characters are likely to involve more or less nonessential so-called terminal reactions. The selection of these for genetic study was perhaps responsible for the now rapidly disappearing belief that genes are concerned only with the control of "superficial" characters. A second difficulty, not unrelated to the first, is that the standard approach to the problem implies the use of characters with visible manifestations. Many such characters involve morphological variations, and these are likely to be based on systems of biochemical reactions so complex as to make analysis exceedingly difficult.

Considerations such as those just outlined have led us to investigate the general problem of the genetic control of developmental and metabolic reactions by reversing the ordinary procedure and, instead of attempting to work out the chemical bases of known genetic characters, to set out to determine if and how genes control known biochemical reactions (Beadle and Tatum 1941b, pp. 499–500, notes omitted).

Throughout the scheme [of control of pigment synthesis in *Drosophila* – the best case previously available in animals – RB] we have indicated genes acting through the intermediation of enzymes. In a sense, this is a purely gratuitous assumption, for we have no direct knowledge of the enzyme systems involved. Since, however, we know that in any such system of biological reactions, enzymes must be concerned in the catalysis of the various steps, and since we are convinced by the accumulating evidence that the specificity of genes is of approximately the same order as that of enzymes, we are strongly biased in favor of this assumption (Beadle and Tatum 1941c).

that enzymes are synthesized *de novo*, so a template mechanism was considered the most likely mode of gene action. This position was reinforced by the idea that a template mechanism would explain the connection between the autocatalytic function of genes (i.e., their ability to make copies of themselves) and their heterocatalytic functions (i.e., their ability to make enzymes and other proteins). Box Two illustrates these two stances in contemporary texts.

The hypothesis of a single precursor for multiple enzymes reinforced the idea that genes should be composed of proteins for two reasons: (1) proteins were the only biological substances known to be able to assume the great variety of structures and configurations that would be required for them to serve as templates; and (2) if genes, as proteins, acted by producing conformal changes in other proteins (which Horowitz believes was not Beadle's favored hypothesis; see note 6), the processes of autocatalysis and heterocatalysis would be closely related instances of template action, thus removing the seeming paradox that the major functions of genes required incompatible mechanisms of gene action (see paragraph two of Beadle 1945a, in Box Two). The early challenges to the protein orthodoxy were taken more seriously than some historians recognize. For example, Beadle (1945a) and Wright (1945) both recognized the importance of the Avery group's findings about transformation of pneumococcus (Avery, MacLeod, and McCarty 1944) as suggesting that, in some cases, nucleic acid might be the crucial component of genes. Beadle handles this point cautiously (Beadle 1945a, pp. 71, 75–6, and 86). For all of these reasons, the one-gene–one-enzyme hypothesis, as first proposed, was conceptually committed to important claims (soon proved to be false) about the nature of gene action, the character of the relevant biochemical interactions, the kinetics of cellular responses to novel carbon sources or antigens, and so forth. This is not surprising, for all the details of gene action were, inevitably, still obscure. To that extent, so were the early formulations of the one-gene–one-enzyme hypothesis (see Box Three).

Still, these false commitments were part and parcel of the very formulation of the hypothesis. To a large extent, the ability to overcome these "mistakes" depended on the search for integration among biochemical, cytological, and genetic knowledge; it was this search that guided the major steps from this point forward. The struggle to reconcile biochemists' and geneticists' accounts of the mechanisms of enzyme synthesis and the controls that governed them yielded much of the evidence on which doctrines concerning the mechanisms of gene action eventually came to rest. It is necessary to understand this struggle (and much more) in detail if one is to understand the paths by which it came to be recognized that nucleotide sequence determines amino-acid sequence in the formation of proteins.

BOX TWO

*Grounds for Supposing that Genes Are Proteins and that Enzyme Synthesis Involves Conformal Change of Precursors*

. . . . either the gene molecules are basic proteins held in the chromonema by longitudinal nucleic acid molecules or they are nucleoproteins some of whose nucleic acid straddle from gene to gene, or they are nucleoproteins alternating with a basic protein-filling substance, to which they are bound on both sides by their nucleic acid valencies (Gulick 1938, p. 163).

Nucleic acid is itself too simple a material and too uniform in nature to be responsible for the specificity of the genes. Similar statements have been made with respect to the protamines and histones extracted from sperms. The possibilities of diversity among protein molecules through the different possible arrangements of the amino acids and through attachment of prosthetic groups is, however, so nearly infinite that there seems to be no theoretical difficulty in connection with gene specificity, even if only a minute portion of the visible chromosome is genic (Wright 1941, pp. 492–3).

This control [of specific biochemical reactions] can be most simply explained by supposing that genes act directly by determining the specificities of enzymes which in turn control the specific biochemical reactions involved [p. 27] . . . If each biosynthesis involves, as seems likely, a series of reactions, a number of genes should be concerned with each synthesis, and it should be possible to block a given process in any one of a number of different places. In several instances we have obtained strains of independent origins in which a given synthesis is defective. In some cases these are genetically and physiologically identical, but in others the two mutations are different genetically and grow normally as heterocaryons [i.e., when multiple nuclei are present within one cell, each bearing one of the mutations – RB]. Presumably in these the synthesis is blocked at different steps in the two mutants. . . . If all the genes concerned in a given synthesis are mutable[,] we should then also be able to estimate the number of steps involved. It should then also be possible to determine the course of the biosynthesis and the type of reaction controlled by each gene by the isolation of intermediate products (Tatum and Beadle 1942, p. 33).

One gland tissue produces one enzyme, another gland produces some other. Is this because in the different tissues different genes are permitted to "do their trick," or is it that different cytoplasms induce the same gene to yield a slightly different product? One is tempted to suppose that each gene produces just one active principle [reference to Tatum and Beadle 1942], and that the differing outcome in the diverse tissues of the body is because this first active product induces more or less different results when it works on different cytoplasmic substrates. This *a priori* assumption may, or may not, be true. It is conceivable [,] even if not too plausible, that feeding a different substrate to the gene may induce it to alter the nature of its own primary product...

The gene's primary products are unknown to us. For the most part, the multitudinous enzymes we find abundantly present in tissues and organs are doubtless not the ones that were produced by genes. The list of enzymes found thus far in the nucleus is not impressive and, except for the nuclear phosphates, they are in rather scant concentration. Seemingly what the genes produce are formative enzymes, doubtless largely of the sort involved in the synthesis of protein molecules, not much known to science, representing early links in the chains of causation that lead up to the later synthesis in the cytoplasm of our more familiar cytoplasmic enzymes (Gulick 1944, p. 19).

Assuming that it is true that every biochemical reaction has a specific gene directing its course, how is this control achieved?...

Genes are thought from various lines of evidence to be composed of nucleoproteins or at least to contain nucleoproteins as essential components. They have the ability to duplicate themselves which, of course, they do once every cell division. The manner in which this self-duplication is brought about is one of biology's unsolved problems, but it is thought to involve a kind of model-copy mechanism by which the gene directs the putting together of the component parts of daughter genes. If this is the mechanism and genes contain protein components, gene reproduction is a special case of protein synthesis. Since many cases are known in which the specificities of antigens and enzymes appear to bear a direct relation to gene specificities, it seems reasonable to suppose that the gene's primary and possibly sole function is in directing the final configurations of protein molecules.

Assuming that each specific protein of the organism has its unique configuration copied from that of a gene, it should follow that every enzyme whose specificity depends on a protein should be subject to modification or inactivation through gene mutation. This would, of course, mean that the reaction normally catalyzed by the enzyme in question would either have its rate or products modified or be blocked entirely (Beadle 1945c, p. 660) [references omitted].

To suppose that a highly specific giant nucleoprotein molecule is formed from nutrients by such a step-by-step process [as is typical of biochemical transformations] seems intolerably complex, even making due allowance for repetitions and for the possibility of catalytic combinations of successively larger blocks. I think that most geneticists have long agreed that there must be shortcut. The gene must somehow act as a model on which a daughter gene is formed as a whole (Wright 1945, p. 293). [On p. 295, the parallel claim is made, without explicit argument, for the mechanism of protein synthesis.]

## A PRELIMINARY MODEL

It *is* fair to accuse me of bias in choosing my case. If ever a problem called for integrating hitherto separate disciplinary treatments of a common subject, it was Beadle and Tatum's problem: genetic control of enzyme production in *Neurospora*. Still, the case is typical of a rather large class of cases. First, it starts from poorly dovetailed descriptions of ongoing processes in particular organisms – in this instance, *Neurospora*, which produce spores that do or do not grow in precisely identifiable circumstances. Second, an account of the details involved in the focal problem must employ descriptors from different disciplines. Here, the descriptors are genetic, cytological, and biochemical: genetic to identify the strains and the genes whose action is to be discovered, cytological to describe the characteristics of the cells and colonies produced, and biochemical to describe the media on which growth is achieved or blocked and the products that accumulate in nutritionally deficient *Neurospora*. Third, as Beadle and Tatum explicitly recognized, *intra*disciplinary approaches to the basic problem could not get around the technical and conceptual difficulties inherent in the separate practices of the three disciplines most centrally involved.

But, it is not obvious how seriously this choice of example affects the normative issues. Indeed, Beadle and Tatum's strategy (described in paragraph two of Box One) is basically derived from that employed in Beadle and Ephrussi's earlier work, which Ephrussi, at least, considered a *failed* attempt to integrate *embryology* and genetics.[7] Ephrussi's aim in that work was to

[7] Compare the account of that strategy offered in Ephrussi and Beadle (1935, p. 98). See Burian, Gayon, and Zallen (1988, pp. 389–400) for a discussion of that work – especially pages 393–5 and 397–8 – for an account of the earlier strategy. Horowitz (personal communication, January 28, 1992) argues that the account presented herein overstates the difference of goals between Beadle and Ephrussi and Beadle and Tatum. From the side of Ephrussi, at least, I disagree; see Burian, Gayon, and Zallen (1988, 1991) for further details.

BOX THREE

*Early Formulations of the One-Gene–One-Enzyme Hypothesis*

. . . [A variety of investigations have] established that many biochemical reactions are in fact controlled in specific ways by specific genes. Furthermore, investigations of this type tend to support the assumption that gene and enzyme specification are of the same order [pp. 499–500] . . . The preliminary results [concerning *Neurospora*] summarized above appear to us to indicate that the approach outlined may offer considerable promise as a method of learning more about how genes regulate development and function. For example, it should be possible, by finding a number of mutants unable to carry out a particular step in a given synthesis, to determine whether only one gene is ordinarily concerned with the immediate regulation of a given specific chemical reaction (Beadle and Tatum 1941a, p. 505).

Such a view does not mean that genes directly "make" proteins. Regardless of precisely how proteins are synthesized, and from what component parts, these parts must themselves be synthesized by reactions which are enzymatically catalyzed and which in turn depend on the functioning of many genes. Thus, to the synthesis of a single protein molecule, probably at least several hundred different genes contribute. But the final molecule corresponds to only one of them and this is the gene we visualize as being in primary control (Beadle 1945c, p. 660) [references omitted]. [See also Beadle 1945a, pp. 18–19, 60, 63, 82, and 86–7; 1945b, pp. 190–2.]

In conclusion, the results so far obtained with *Neurospora* support the synthesis that genes concerned in biosyntheses, and probably all genes, act in a primary way by determining the specificity of or in controlling the production of enzymes. The results also support the view that a one-to-one relation exists between gene and enzyme. At present it seems likely that any apparently multiple-gene effects in *Neurospora*, when completely analyzed biochemically and genetically, will be found to be due to common primary reactions, or to secondary interactions not directly related to the action of the mutant gene under consideration (Tatum and Beadle 1945, p. 129).

perform genetic analysis on characters at the end of the developmental chain instead of starting from genes and analyzing whatever characters those genes were found to determine. The reasons for Beadle and Ephrussi's failure to achieve the integration they sought (which is *not* the same thing as failure of their research!) are precisely those described in paragraph one of Box One. Their inability to overcome the complexities of the developmental pathways of *Drosophila* should not be considered a serious argument against the soundness of the underlying strategy that they employed. Rather, it shows that successful problem solving requires sufficiency of the available means.

In both cases, the need to bring multiple modes of description to bear on the problem of concern (i.e., genetic control of enzyme synthesis for Beadle and Tatum; genetic and developmental control of pigment synthesis in a higher organism for Beadle and Ephrussi) forced the program of research to assume an interdisciplinary character. And, the internal dynamic of the disciplines in question produced some seemingly paradoxical results (e.g., for Beadle and Tatum, regarding the kinetics of enzyme synthesis and the relations among the components of nucleoproteins). In both cases, it became imperative to seek a coherent account of the competing descriptions provided by the disciplines involved, laden with discipline-specific presuppositions and theoretical commitments.

Such seeming paradoxes are at least as likely to arise in cases that do not yield successful integrations as in those that do. Indeed, a central reason for considering the one-gene–one-enzyme hypothesis to be successful is that the positions taken initially by the principals and by the vast number of research scientists who came to employ the hypothesis were transformed by the demand for coherence and the insistent and continued operation of the norm of integration. The resultant transformations in beliefs and practices contributed to a reconciliation of many of the conflicting claims propounded by biochemists, cytologists, and geneticists regarding the distribution and function of materials in the cell, the sequence of events in enzyme synthesis, the relationships among nuclear and cytoplasmic factors in determining specificity, and the means by which to evaluate differences regarding such issues. But these transformations also complicated the knowledge of gene action to the point that, by the end of the 1950s, the original hypothesis no longer was considered literally true and no longer played a central role in organizing the strategy of research.

All of this suggests a preliminary moral: *Once scientists recognize that their claims about a common subject matter conflict, they face a higher order norm according to which they ought to (attempt to) reconcile their differences.* It is worth noting that the conflicting claims can be at any level. The arguments between Darwinians and the followers of Kelvin regarding the age of the earth demonstrate that the issue can be over a "simple matter of fact" as easily as it can be over the theoretical or conceptual commitments employed in describing or explaining a range of phenomena.

The means for resolving the conflict may not be at hand; resolution may have to await the invention of new techniques or the adaptation of techniques borrowed from elsewhere. In retrospect (especially in light of Beadle and Tatum's success), it is clear that Beadle and Ephrussi did not have the tools they needed to succeed in their attempt to integrate embryology and genetics or to provide a genetic account of developmental pathways. Nonetheless, the need for reconciliation remains obvious in such cases – and it brings with it the obligation to develop the tools required for resolution. Such situations suggest many difficult questions about the matching of tools, organisms, institutions, and conceptual frameworks required to solve major problems. These are among the questions that, I believe, will play a central role in the next decade of work in history, philosophy, and (yes!) sociology of biology.

We have here a small beginning of an account of when one ought to seek integration and of when it might count against one's work that the quest for integration had failed. It is an account that depends, critically, on the central role of the norm of a certain sort of disciplinary integration as an objective of science.

## NORMATIVE METHODOLOGY

The *standard of progress* implicit here is reasonably clear: It is the achievement of concordance among discrepant discipline-based accounts of pertinent phenomena. This standard rests on the desirability of integration; it assumes, in turn, that (for properly formulated questions) there is but one (multiply describable) truth. However, this standard of progress, by itself, tells us nothing about the circumstances in which integration is *feasible*, nor about the form it should take. The difficulty of determining feasibility provides ample grounds for skepticism regarding the value of insistence on integration in the short term, thus reinforcing some of van der Steen's (1993) concerns.

However, the openness of the *form* of integration raises a difficulty for one of van der Steen's principal criticisms of integration as a norm. It also poses problems for Schaffner's General Reduction–Replacement Model (GRR) insofar as the latter aims to provide significant methodological guidance to scientists.

Let us take these difficulties one at a time. Van der Steen (1993) argues that the combinatorics of integrating all pairs of disciplines are prohibitive. But this combinatorial argument has a dubious premise – to wit, that successful integration proceeds pairwise between arbitrary disciplines rather than between each of many "peripheral" disciplines and one "central" discipline.

One virtue of old-fashioned reductionism was that it favored the integration of *many* theories or disciplines by means of a single grounding theory or under the umbrella of an overarching discipline. Van der Steen's combinatorial argument is irrelevant where such a thing is possible. Even if one is pessimistic about anything resembling full-fledged reduction, one need not envisage separate integration between every pair of disciplines.

Indeed, such a picture is absurd. Consider the sequel to the work on the one-gene–one-enzyme hypothesis. That work, plus much related work, belongs near the center of a set of problems, techniques, and results that ultimately provided the basis for a new discipline: molecular genetics. Over the course of about thirty years, the development of that new discipline reformed a massive region of the disciplinary and intellectual landscape in biology in such a way that genetics became inescapably relevant to problems that had, until recently, been impervious to genetic considerations. Molecular genetics gradually came to provide the techniques of choice for dealing with hitherto intractable problems. Genetic knowledge and techniques came to be applied to an evolving and expanding domain of problems, but there was nothing like the spawning of two hundred new disciplines. With time, genetic techniques became the primary tools, and knowledge of the mechanics of gene action and gene regulation provided the primary insights, for a large number of basic problems. "Borrowings" became ever more frequent. This, rather than pairwise formation of hybrid disciplines, was the primary mode of unification.

Turning to Schaffner's GRR (Schaffner 1993a; see also Schaffner 1993b), the difficulty is this: Suppose one can find reduction or replacement connections satisfying the GRR model *after* a successful integration. Unless one has independent grounds for specifying in advance both the *form* (reduction or replacement?) and the *direction* in which to go, the model can provide only *ex post facto* justification of a scientist's research strategy. More to the point, it can justify incompatible strategies equally well, depending on which discipline one takes as the starting point.

Beyond this general criticism of the specificity of the GRR model, I have substantive concerns about its adequacy. Unlike Schaffner, I hold that the most fruitful reconstruction of molecular biology does not focus on a central theory in the classical philosopher's sense of a body of laws from which, by use of initial or boundary conditions, one derives an account of what to expect in particular cases. Molecular biology is more like (theoretical!) automobile mechanics than it is like Newtonian mechanics; it is best understood as detailed knowledge of and a body of techniques for studying a congeries of molecular mechanisms whose operation depends only on *general* physico–chemical laws. It is by means of these mechanisms that information is processed and transmitted, that substances in physiological contexts enter into chain reactions to yield particular products and alter morphology, and so on. Burian (1996) offers an argument in favor of such an account.

This criticism illustrates the limited resources available to normative methodology. A rational reconstruction along the lines of GRR is useful for analyzing theory relations after the fact, but it does not usefully discriminate among alternative strategies for those who actively seek to forge relations between theories or disciplines in conflict. For this purpose, the best that we can offer is less formal *higher order norms*, operating at a meta level or functioning as rules of thumb. Such norms say something about *what is yet to be done* but little about *how to do it*.

One reason that we cannot do better is that the professionalization of scientists within different disciplines equips them with distinctive practices, standards for evaluating knowledge claims, and disciplinary norms for generating and evaluating knowledge, with no generally satisfactory means of resolving the conflicts that occasionally result.[8] Scientists cannot escape the necessity of working with "local" standards in this sense. This is illustrated by Bechtel's analysis (1993) of the formation of the discipline of cell biology, with its own ethnocentrism. This *inevitable ethnocentrism* means that progress should be measured as much in terms of the changing interrelations of disciplines as in terms of the particular advances made by particular scientists within a discipline or a "cross-disciplinary research cluster."

Van der Steen, therefore, is right to insist that there are many competing norms in science and that one of the difficulties the scientist faces is how seriously to take which norms in which circumstances. In a genuinely pluralistic enterprise, no substantive norm or group of norms can characterize all of the work that is undertaken at a single time. In this respect, the norms of the separate disciplines – and their terminologies, presuppositions, weightings of evidence, and so on – should be allowed to operate with considerable autonomy and without much mutual interference *in the short term*. But when work in different disciplines bears on a given problem, their practices, terminologies, and standards for evaluation of experimental evidence must be brought into accord with respect to the matter at hand for it to count as satisfactorily solved.

Thus, *the solution of a major problem requires some form of coherence or unification*. In terms of the case study undertaken here, the claim that genes

[8] The practices, standards, and norms are open-ended and not wholly rigid but subject to revision. The interplay of factual knowledge with disciplinary norms allows results obtained by work in other disciplines eventually to penetrate and alter the practices and standards of a given discipline.

determine the specificity of enzymes cannot stand alone; sooner or later, it must be integrated with other biochemical findings or it will be infirmed by its failure to apply to the chemical reactions that actually take place in real organisms. Similarly, if the study of the detailed structure, behavior, and internal mechanisms of cells is to provide an adequate description of the mechanisms underlying cellular behavior and morphology, it should yield a coherent body of descriptions (whether or not they are unified by a theory in any classical sense).

What do such examples mean for our normative questions? Not that the norms governing genetic, biochemical, cytological, physiological, neurobiological, or evolutionary investigations should be sacrificed to some overarching requirement of unification. Rather, *after investigations in separate disciplines have gone their own way, where those investigations produce conflicting or poorly matching claims about particular cases, about the mechanisms or processes underlying those cases, or about the expectable patterns or laws, the matter cannot be taken as settled. In consequence, scientists must resolve the discrepancies between the claims of different disciplines before the problems on which they focus can count as solved.* In cases of conflict, unification remains a central higher-level norm that, even if it does not enter the everyday life of particular scientists or laboratories, must be satisfied for there to be a convincing resolution of the issues at hand. The norm of unification requires that such long-range problems not be dropped unless they are satisfactorily eliminated by problem reformulation or transformation.

It is important for the health of science that the work of and standards employed in different disciplines be made mutually coherent. Attempts to achieve coherence often result in what Schaffner focuses on in his case study – to wit, the analysis of causal sequences across levels of aggregation (e.g., biochemical causes of cellular behavior). But such local instances need not and, I predict, often will not write up into globally unified causal mechanisms or theories. What must be made to cohere are the *general causal analyses* of the different disciplines and their interpretations of the results of applying various experimental techniques to particular cases.

When things are well behaved, integration may yield a neat reductionist story; planets behave nearly like Newtonian mass points acting under gravitational attraction, and many gases behave in good accord with Maxwellian thermodynamics. But in cases more typical of the biological sciences, in which the behavior of higher-level entities is effectively supervenient on that of lower-level entities, no broad unification along the lines of an extended theoretical reduction ought to be expected. Even where there are "mechanisms all the way down," our higher-level descriptions may not (and, I think, often will not) map neatly on mechanistic descriptions of the phenomena at lower levels of aggregation. Unification does not require reduction of the sort I am here rejecting, not even as a norm. This is the moral I would draw, in contrast to Schaffner, from his studies of "theories of the middle range."

What weight should be given to unification as a methodological objective? The answer is highly variable and must be justified locally. The justification must pertain to the particular domains and problems in question and must be conducted in light of what is known about the relevant objects and processes and about the techniques used to investigate them. It is a matter of judgment when to take the norm of unification seriously. Van der Steen is right: Premature unification of concepts is harmful; however, this in no way undercuts the legitimacy of unification as a generally applicable higher-level norm.

It is not obvious that a properly developed account of premature unification will turn (as van der Steen expects) on conceptual as opposed to factual considerations. Our case study helps make this point. In retrospect, the fusion that Beadle sought to achieve between biochemical and genetic concepts could not have worked along the lines that he pursued *not* because there was a conceptual error in attempting to analyze the gene as a (nucleo)protein or as a heterocatalytic template, but because as a *matter of fact* genes are not entities of this sort and do not behave in the manner thus predicted. Furthermore, to detect the misbehavior of particular genetically characterized strains (given the techniques available at the time), one had to appeal to biochemical and cytological findings. This meant that one had to employ biochemical standards to characterize *genetic* lesions – enforcing a change in practice that eventually had fundamental consequences for genetics. But if the findings of genetics, cytology, and biochemistry had to be correlated to establish what was and what was not under genetic control, then the factual (and resultant conceptual) errors in Beadle's (partially premature) attempt at unification could best be detected by pursuit of unification as a goal – by attempting to unify the descriptions provided by each of the three disciplines at stake of the behavior of *Neurospora* and other organisms.

On what should the weighting of methodological norms depend? The quickest answer is not very satisfactory: on many things, varying widely from case to case. To make the point in a quick and dirty way, the feasibility of unification depends *inter alia* on the limitations of available techniques. Until one could reliably separate cytoplasmically controlled phenomena from those controlled by nuclear genes in *Neurospora*, experiments on this organism would not yield unification of genetic and biochemical concepts. Until one could reliably bring about genetic exchange between strains of bacteria, limitations of technique severely restricted the degree of unification that could be expected between bacterial physiology and the extensions of Mendelian genetics.

These are, of course, not the only pathways by means of which conceptual links could be forged between studies of microorganisms and the preexisting work in genetics; *however, without some sort of technological advances beyond what was available in the late 1930s, such links could not have been successfully forged.* And this is a factual, not a philosophical, claim.

Typically, as new techniques and ways of exploiting the properties of particular organisms or groups of organisms are found, a series of questions arises regarding the connections with other information about those organisms and the various theoretical descriptions employed in accounting for the behaviors in question. Armed with the right questions, new techniques open up the opportunity to seek unification where it has not been available. And, armed with new concepts – sometimes inspired by empirical findings and sometimes not – scientists revise the list of available and technically tractable questions, thereby altering the rankings of plausibility among theories and hypotheses. Thus, *retaining unification as a long-term objective is the key to exploiting many technical advances so as to resolve hitherto intractable problems.* To argue with van der Steen that the higher-order aim of integration of knowledge is often damaging is to miss its importance as a guiding ideal, *even if one cannot expect a full-scale integrated theory in biology.*

## CONCLUSION

The consequence of these considerations is this: A methodologically central higher-order goal of the biological sciences is unification of theories and descriptions across disciplinary boundaries so as to achieve coherence of descriptions and explanation in application to particular cases. That goal can be mobilized, at best, only where specific considerations suggest that different terminologies that do not map easily onto one another or that yield contradictions when forcibly combined nonetheless overlap in their reference to entities, processes, or mechanisms. Attempts at unification fail more often than not, for the implicit methodology of unification is weak in the short term. Still, it is powerful in the middle or the long term. That methodology consists of the attempt to find explicit linkages of a great variety of types among the entities, properties, behaviors, and processes described in one discourse and those described in another – linkages that will allow the separate modes of discourse to mesh together without necessarily becoming intertranslatable (cf. recent work on supervenience).

To the extent that this account is sound, it is incumbent on us to investigate the factors that support or infirm the judgment that neighboring disciplines are not sufficiently coherent in their descriptions of common problems or in their findings about or explanations of phenomena that fall in both of their respective domains of knowledge. It is also incumbent on us to find ways of estimating the relative importance of bringing neighboring disciplines into line with one another in a given instance as opposed to elaborating the programs of each in independence of the other. This chapter has accomplished its purpose if it has presented a persuasive case to the effect that one ought to be able to articulate an account of the sorts of considerations that enforce a high priority on local (or, better, regional) unification and that the resultant methodology is not wedded to a grand scheme of the unity of science or the traditional pictures of theoretical unification or theory reduction.

## REFERENCES


Astbury, W. T. 1938. "Some recent developments in the X-ray study of proteins and related structures." *Cold Spring Harbor Symposia on Quantitative Biology* 6: 109–21.

Avery, O. T., C. M. MacLeod, and M. McCarty. 1944. "Studies on the chemical nature of the substance-inducing transformation of pneumococcal types. I. Induction of transformation by a deoxyribonucleic acid fraction isolated from pneumococcus type III." *Journal of Experimental Biology and Medicine* 79: 137–58.

Beadle, G. W. 1945a. "Biochemical genetics." *Chemical Reviews* 37: 15–96.

Beadle, G. W. 1945b. "The genetic control of biochemical reactions." *Harvey Lectures* 40: 179–94.

Beadle, G. W. 1945c. "Genetics and metabolism in *Neurospora*." *Physiological Reviews* 25: 643–63.

Beadle, G. W., and E. L. Tatum. 1941a. "Genetic control of developmental reactions." *American Naturalist* 75: 107–16.

Beadle, G. W., and E. L. Tatum. 1941b. "Genetic control of biochemical reactions in *Neurospora*." *Proceedings of the National Academy of Sciences, USA* 27: 499–506.

Beadle, G. W., and E. L. Tatum. 1941c. "Experimental control of development and differentiation: Genetic control of developmental reactions." *The American Naturalist* 75: 107–16.

Bechtel, W. 1993. "Integrating sciences by creating new disciplines: The case of cell biology." *Biology and Philosophy* 8: 277–300.

Burian, R. M. 1996. "Underappreciated pathways toward molecular genetics as illustrated by Jean Brachet's chemical embryology." In *The Philosophy and History of Molecular Biology: New Perspectives*, ed. S. Sarkar. Dordrecht, Holland: Kluwer, 67–85.

Burian, R. M., J. Gayon, and D. Zallen. 1988. "The singular fate of genetics in the history of French biology, 1900–1940." *Journal of the History of Biology* 21: 357–402.

Burian, R. M., J. Gayon, and D. Zallen. 1991. "Boris Ephrussi and the synthesis of genetics and embryology." In *A Conceptual History of Embryology*, ed. S. Gilbert. New York: Plenum, 207–27.

Darden, L. 1991. *Theory Change in Science: Strategies from Mendelian Genetics*. New York: Oxford University Press.

Ephrussi, B., and G. W. Beadle, 1935. "La transplantation des disques imaginaux chez la *Drosophile*." *Comptes rendus de l'Académie des Sciences, Paris* 201: 98–100.

Gulick, A. 1938. "What are the new genes? II. The physico–chemical picture. Conclusions." *Quarterly Review of Biology*.

Gulick, A. 1944. "The chemical formulation of gene structure and gene action." *Advances in Enzymology* 4: 1–33.

Haeckel, Ernst. 1876. *Natural History of Creation*. New York: Norton.

Kay, L. E. 1992. *The Molecular Vision of Life: Caltech, the Rockefeller Foundation, and the Rise of the New Biology*. New York: Oxford University Press.

Maynard Smith, J., R. M. Burian, S. A. Kauffman, P. Alberch, J. H. Campbell, B. Goodwin, R. Lande, D. M. Raup, and L. Wolpert. 1985. "Developmental constraints and evolution: A perspective from the Mountain Lake conference on development and evolution." *Quarterly Review of Biology* 60: 265–87.

Monod, J. L. 1947. "The phenomenon of enzymatic adaptation and its bearing on problems of genetics and cellular differentiation." *Growth Symposium* 11: 223–89.

Ruse, M., and R. M. Burian. 1993. "Integration in biology, a special issue." *Biology and Philosophy* 8, no. 3.

Schaffner, K. F. 1993a. "Theory structure, reduction, and disciplinary integration in biology." *Biology and Philosophy* 8: 319–47.

Schaffner, K. F. 1993b. *Discovery and Explanation in Biology and Medicine*. Chicago: University of Chicago Press.

Spiegelman, S. 1950. "Modern aspects of enzymatic adaptation." In *The Enzymes*, eds. J. B. Sumner and K. Myrback. New York: Academic Press, 267–306.

Tatum, E. L., and G. W. Beadle. 1942. "Genetic control of biochemical reactions in *Neurospora*: An 'aminobenzoicless' mutant." *Proceedings of the National Academy of Sciences, USA* 28: 234–43.

Tatum, E. L., and G. W. Beadle. 1945. "Biochemical genetics of *Neurospora*." *Annals of the Missouri Botanical Gardens* 32: 125–9.

van der Steen, W. J. 1990. "Concepts in biology: A survey of practical methodological principles." *Journal of Theoretical Biology* 143: 383–403.

van der Steen, W. J. 1993. "Toward disciplinary disintegration in biology." *Biology and Philosophy* 8: 259–76.

van der Steen, W. J., and P. B. Sloep. 1988. "Mere generality is not enough." *Biology and Philosophy* 3: 217–19.

van der Steen, W. J., and P. J. Thung. 1988. *Faces of Medicine: A Philosophical Study*. Dordrecht, Holland: Kluwer.

Wright, S. 1941. "The physiology of the gene." *Physiological Reviews* 21: 487–527.

Wright, S. 1945. "Genes as physiological agents." *American Naturalist* 79: 289–303.

# II

# Evolution

The three chapters in this section were originally presented to quite different audiences. Chapter 4, on concepts of adaptation, was first presented at a 1981 workshop organized by Marjorie Grene concerning issues about Darwinian evolutionary theory; it was published in the volume that resulted from the workshop (Grene 1983). The participants were all advanced professionals in biology, history of biology, or philosophy of biology, so I do not set as much background as I do in the other chapters of this section. (Indeed, readers without prior background in evolutionary biology may wish to read Chapters 5 and 6 before Chapter 4.) Chapter 4 teases apart several concepts of adaptation and fitness deployed by Darwin and in the synthetic theory of evolution, which achieved orthodoxy around 1950. These concepts are often conflated with one another and need to be carefully separated to achieve a sharp understanding of the claims of various evolutionary theories. The analysis reveals some important substantive differences between Darwin's theory and the synthetic theory. In addition, it shows how easy it is to be misled into believing that the synthetic theory rests on a tautology – and does so by providing a careful account of exactly why survival of the fittest is not properly interpreted, tautologously, as survival of those that actually survive. It also offers some guidance in handling a number of delicate questions about the strength of the evidence for claims that particular traits are evolutionary adaptations by providing an account of requirements that must be met for the *actual* survival of organisms to provide evidence about the contribution (if any) that possession of particular traits makes to the fitness, adaptedness, or adaptation of the organisms that possess them. Chapter 4 thus illustrates the contribution of conceptual analysis to the interpretation of evidence, thereby demonstrating, I believe, that conceptual clarity can be important for working scientists even though, by itself, it is not enough to solve biological problems.

Chapter 5 was originally presented at a 1987 symposium on "Evolutionary Biology at the Crossroads." The symposium was organized by Max Hecht, a distinguished evolutionary biologist, as part of a year-long celebration of the fiftieth anniversary of Queens College of the City College of New York and of Theodosius Dobzhansky's lectures at Columbia that resulted in *Genetics and the Origin of Species* (Dobzhansky 1937). I was asked to address both the wider audience and the many experts who participated in this festive occasion. In keeping with the assigned topic, the chapter explores the respects in which there is (and there is not) an evolutionary paradigm, some important features of evolutionary biology and evolutionary theory, the interpretation of the synthetic theory of evolution, and the near-term prospects for developments in evolutionary biology. I argue that organismal (in contrast to stellar) evolution is a deeply historical process in which the contingencies of circumstance and interactions among individuals and across levels of organization cannot be ignored. In consequence, evolutionary biology cannot be satisfactorily grounded in a fundamental theory as such but must be built on cross-disciplinary interchange in which the various influences on evolution, and their interactions, are taken into account. Chapter 5 connects these matters to the history of Darwinism and evolutionary biology, arguing that the variation in the interpretation of Darwinism and of the synthetic theory of evolution (and the pluralism of evolutionary biology) are expectable consequences of the lack of a single paradigm, in the sense of a disciplinary matrix, in evolutionary biology.

Chapter 6 derives from a paper presented at a conference at the University of Burgundy in honor of Marjorie Grene in 1995. In part because of the venue and in part because the audience was quite diverse, I focus on an article by Dobzhansky, aimed at biology teachers, on the centrality of evolution in biology. Dobzhansky's article provides an opportunity for locating evolutionary issues in a somewhat larger cultural matrix, including some differences in the reception of evolution in different countries and some peculiarities of its reception in the United States. I also discuss Dobzhansky's arguments about the evidential status of evolution, the place of evolution within biology, and the importance of selection to the course of evolution. These matters lead to discussion of the evidential issues raised by Dobzhansky, the importance of contingency in evolution, and the relationships among biological disciplines. Chapter 6 also begins to open up a topic that is more important in subsequent chapters of this book – to wit, the impact of molecular biology on our understanding of evolution and on the relationships among biological disciplines.

## REFERENCES

Dobzhansky, T. 1937. *Genetics and the Origin of Species*. New York: Columbia University Press.

Grene, M., ed. 1983. *Dimensions of Darwinism: Themes and Counterthemes in Twentieth-Century Evolutionary Theory*. Cambridge: Cambridge University Press.

# 4

# "Adaptation"[1]

In this chapter, I examine three concepts of adaptedness and adaptation found in Darwin and two additional concepts of "Darwinian fitness" employed in current population biology and evolutionary theory. The examination reveals a number of conceptual confusions within evolutionary biology and shows that much of the philosophical literature about evolutionary theory has been concerned with vulgarizations of that theory. Several recent controversies in evolutionary theory, all of which focus on the degree of control that natural selection exercises over traits of organisms, are mentioned in passing. These controversies have been plagued by confusions regarding what it is for a trait to be an adaptation, the kinds of evidence required to support such a claim, and the units of analysis appropriate to applications of the concept of (an) adaptation. Removal of these confusions may contribute to the attempt to resolve the underlying biological disagreements about the relative importance of various units of selection, the existence of neutral mutations, "selfish" DNA, and so forth, but it will not be sufficient, by itself, to resolve those disagreements. Removal of those confusions is also an important step toward a deeper and more fruitful understanding of the structure and content of evolutionary theory.

[1] Originally published in Marjorie Grene, editor, *Dimensions of Darwinism* (New York and Cambridge: Cambridge University Press, 1983), pp. 287–314. Reprinted with kind permission of Cambridge University Press. This chapter has benefited from criticisms by F. Ayala, C. Bajema, R. Brandon, L. Darden, S. Gould, M. Grene, K. Guyot, Ph. Kitcher, E. Mayr, R. Richardson, P. Richerson, A. Rosenberg, G. Simmons, and many others, including discussants at the Reimers Stiftung, Bad Homburg; the Committee on History and Philosophy of Science of the University of Maryland; and the seminar on History and Philosophy of Science, University of California, Davis. I am grateful to all concerned.

## ADAPTATION AND ADAPTEDNESS

In a book on *The Development of Darwin's Theory* of evolution by natural selection, the late Dov Ospovat (1981) argued, in effect, that a major key to understanding Darwin and his theory is provided by grasping the changes in Darwin's concepts of "adaptation" and "adaptedness" and the subsequent changes that Darwin's work helped to bring about in the use of cognate concepts by biologists. Although I am not entirely persuaded by all the details of Ospovat's account of Darwin's concepts of adaptation,[2] I employ some of his views, uncritically, as a starting point. The issues about the content and character of the concept of adaptation that I explore will not be greatly altered (though they may be clarified) by further historical scholarship regarding the path that Darwin followed.

Before proceeding, I must address one terminological point. As will become apparent in course of this chapter, the use of such terms as "fit," "adapted," "fitness," and "adaptation" in the biological and philosophical literatures has been confused and confusing. To build my account of the relevant concepts, I have found it useful to fix one aspect of usage by fiat. I employ the term "adaptation" in a historical sense, the terms "adapted" and "adaptedness" in an *a*historical sense. Fleetness contributes to the adaptedness of a deer (or makes the deer better adapted) if and only if, other things being equal, it contributes to the solution of a problem posed to the deer – for example, escaping predation. Fleetness is an adaptation of the deer if and only if the deer's fleetness has been molded by a historical process in which relative fleetness of earlier deer helped shape the fleetness of current deer. This usage (which, in its present sharp form, derives from a suggestion by Robert Brandon) is my own, though it reflects an important tradition in the literature. This specialized terminology facilitates the clarification of a number of the confusions to be discussed subsequently.

Ospovat documents Darwin's struggle to arrive at a satisfactory concept of adaptation for use in the theory of evolution by natural selection. According to Ospovat, Darwin first employed two distinct notions of "absolute" or "perfect" adaptedness (only one of which I discuss) in the period from 1838 to 1854 and then, only after considerable difficulty, arrived at a version of the notion of relative adaptedness employed in the *Origin* sometime in 1857. However Darwin got to it, Ospovat is surely right that the notion of relative adaptedness developed by Darwin is of crucial importance to evolutionary

[2] Nor of some of his chronology regarding the development of Darwin's thought.

theory. Yet, I am prepared to argue (1 can give only part of the argument herein) that the development of the modern neo-Darwinian (or synthetic) theory of evolution has, in fact, minimized the role of Darwin's concepts of adaptation and adaptedness and obscured their importance by running them together, in a confused and confusing fashion, with a whole battery of other related notions. For instance, two quite un-Darwinian notions, both of which pass under the label of "Darwinian fitness," are common in population biology and evolutionary theory. The widespread use, and occasional misuse, of these notions has, at times, contributed to the illusion that evolutionary theory is viciously circular or tautologous. Distinguishing between these concepts of Darwinian fitness (so-called) and related concepts of adaptation, including Darwin's, is crucial to the proper understanding of evolutionary theory.

Until about 1975, when a number of new papers bearing on these topics began appearing, the most important and influential attempt to straighten out the conceptual confusions involved in various recent uses and misuses of terms like "adaptation" and "fitness" was G. C. Williams's *Adaptation and Natural Selection* (Williams 1966). Williams's main concern was to bring about correct use of the concept (or concepts) of adaptation in contemporary biology – not by means of a logic-chopping analysis but rather by showing biologists how practical and theoretical difficulties arose from their misuses of the relevant concepts and by persuasive arguments to the effect that the biological difficulties and disagreements could be handled most effectively by cautious and correct use of the concept of adaptation along the lines he advocated. Williams's book was an immensely useful corrective to the near loss of Darwin's concept in theoretical biology. But it is now time to correct certain flaws in Williams's treatment of the concept of adaptation. In particular, his account of adaptation is wedded to quite strong and controversial commitments about the effectiveness of natural selection and about the units of selection – that is, the level at which selection is most effective. I hope to separate the concept of adaptation from these commitments.

I proceed indirectly, carrying out only part of this rather large task. I try to fill in certain parts of the relevant background, mainly by developing my own account of some of the pertinent concepts of adaptation, fitness, and adaptedness, and discussing their interrelations and the confusing consequences of their interplay. In a sequel to this essay,[3] I show in detail the relevance of these

[3] See my "On some controversies concerning macroevolution" (unpublished MS, never completed).

matters to contemporary concerns about the units of selection and exhibit the interplay between controversies concerning group selection, macroevolution, and related topics. In this chapter, my main target is to carry the account of the concept of adaptation a few steps beyond Williams.

Before proceeding with this agenda, I examine briefly what we should expect of an account of the concept of adaptation. How can we tell a good account from a bad one?

If, as I argue, many biologists are careless and inconsistent in their use of "adaptation," "fitness," and cognate terms, we should not make our evaluation of an analysis of the concept of adaptation turn on mere conformity to usage. Rather, it seems to me, we must isolate certain central claims to serve as touchstones against which to test alternative analyses. As the analysis becomes more sophisticated, we may have to reexamine and improve the touchstones, to divide and complicate our account of the concepts involved.

To illustrate what I mean by a touchstone, consider a thought experiment tracing back at least as far as Scriven (1959). In this experiment, two identical twins are standing in a forest during a thunderstorm. One is struck by lightning and killed, the other is unhurt; he later marries and has ten children. One may now ask whether one twin is *better adapted* (or *fitter*) than the other. In conformity with *Darwin's* usage – which, I maintain, is standard in this regard – the twins have the same degree of adaptedness because the difference between them is a matter of happenstance and they are, in all biologically relevant properties, virtually identical; they are equally fit or unfit vis-à-vis their environment. Later, we will see that use of this little fable as a touchstone shows that one of the most common measures of so-called Darwinian fitness in population biology does not serve as a direct measure of fitness in the sense of adaptation. But I save that point, as well as two further touchstones, for the appropriate juncture of the argument.

## DARWIN ON ADAPTATION

Let me start the real work of this chapter by turning to Darwin's approach to adaptation. On Ospovat's account of Darwin's early unpublished work bearing on evolution by natural selection, Darwin held, at least from 1838 to 1850, a theory quite different than the one for which he is known. A remarkable feature of this theory is that it was based on the concept of *limited perfect adaptedness*. According to this theory, organisms are constrained by their structure and constitution – by being built out of organic, not inorganic, matter; by being structured as vertebrates or invertebrates; by being warm-blooded or cold-blooded; and so on. Within the limitations of the constraints thus imposed, each species was – or was normally – *perfectly adapted* to its environment. By this was meant that the organisms in question were optimally designed within the applicable constraints to solve the problems posed by the environment: heat, cold, wind, rain, opening up seeds, capturing prey of the sizes and speeds available, and so on.

One of Darwin's central concerns was to explain the diversity of living organisms. At this phase of his development, according to Ospovat, he held that diversity is a secondary consequence of *three* factors:

1. Gradual geological change in general, which altered the environment slowly and imperceptibly but which ultimately confronted organisms with environments to which they were not perfectly adapted.
2. Isolation of one population from another, meaning that different populations of organisms belonging to the same species would face different environments and, hence, different design problems.
3. The laws of variation (what Darwin in the *Origin* called "the mysterious laws of variation and correlation of parts") which, at this time, Darwin thought were such as to call forth variation *only when adaptedness became imperfect*, and then to call forth *directed* variation – variation adjusting the properties of the organism in the direction of perfect adaptedness vis-à-vis the new environmental circumstances.[4]

The role of natural selection at this stage in the development of Darwin's theory, if Ospovat is correct, was dual: First, it guaranteed that there would be pressure on all populations to maintain or to achieve perfect adaptedness; and, second, it preserved the most favorable variations and moved the population mean toward them. Variation provided the *direction* of evolution, whereas the *reduced* variation of perfectly adapted organisms ensured that natural selection would alter the constitution of a population only when the organisms in question were not perfectly adapted.

Whether or not this thumbnail sketch, or Ospovat's detailed account, is fair to Darwin's private speculations, it points to certain features of his concept of adaptedness, of his account of what it is for an organism to be fit with respect to its environment, common to the notebooks and early drafts and to the *Origin*. I capitalize on these: To put the essential points informally, an organism is well adapted when its structure and habits enable it to solve expectable challenges of the environment optimally. An organism is fit or

[4] We see here that Darwin already had good reasons for taking the inheritance of acquired characters seriously in his earliest speculations on transformation; such inheritance might provide a mechanism that would give variation its direction.

unfit according to its ability to meet these challenges by virtue of its design and its programmed patterns of behavior. Employing engineering criteria, it is possible to compare organisms by type and to determine that organisms of type 1 are adapted (designed) to meet challenges A (or challenges A to Z), whereas organisms of type 2 are not adapted (or not as well adapted) to meet that challenge (or those challenges). Certain features of organisms – for example, "organs of extreme perfection" like the vertebrate eye – can be recognized as adaptations; that is, as features existing in some sense or other *because* they are designed for – adapted to – the performance of certain tasks useful or necessary for the survival of the organism.

One reason for citing Ospovat's account of Darwin's early work is that it highlights the possibility of having both absolute and relative concepts of adaptedness. An absolute concept evaluates the design of a product – for example, the eye or the whole organism – in its own right. To evaluate the design, it is necessary to specify, tacitly or explicitly, the design problem and the pertinent constraints governing its solution; from there, evaluation can proceed by some form of "static" engineering analysis. Because the analysis is static, not dynamic, the absolute concept of adaptedness has no direct bearing on the process by which the design was achieved. Accordingly, it needs supplementation if it is to play a role in the theory of evolution. Darwin's early theory provided that supplementation by reference to "the mysterious laws of variation" that, he thought, directed heritable change toward optimal design whenever an organism was insufficiently adapted. In this theory, natural selection played a relatively secondary role – it was the force that called unadapted organisms to account and it was the preserver of adaptations – that is, of optimally designed features and organisms.

Once Darwin came to realize that variation is ubiquitous and largely undirected with respect to the needs of the organism,[5] he was forced to employ a relative concept of adaptedness, a concept tied much more intimately to the process of natural selection than the absolute one was. All organisms face a multitude of problems bearing on survival and reproduction. If they all vary (at least slightly) in virtually all their features, then typical organisms are not perfectly adapted. Some, however, are at an advantage with respect to

[5] P. Richerson (University of California, Davis) reminded me that one must not lose sight of the fact that Darwin retained some role for directed variation throughout his career. The supposed effects on an organism's descendants of its use or disuse of its organs, amplified in various "neo-Lamarckian" ways in Darwin's later theory of pangenesis, mark the strength of Darwin s commitment to directed variation. In his later work, he employed directed variation to speed up evolutionary processes and to overcome the difficulties posed by (theories of) blending inheritance (see Lord Kelvin's and F. Jenkin's criticisms of Darwin's views). For a brief exposition regarding directed variation, see Darwin (1859) and Gould (1971).

others – that is, they are better designed to meet the expected or expectable challenges of the environment. These organisms would, therefore, be more likely to survive the insults of the environment. To be sure, their design offers no guarantee that they will do so, but it provides a statistical bias in their favor. To the extent that two further conditions are met, *natural selection is likely to increase adaptedness*: first, that the environment is stable enough that the problems organisms face are reasonably constant or predictable;[6] and, second, that the offspring of an organism or of a pair of organisms tends to differ from the population mean in the same ways that its parents do. (This latter condition, the "statistical heritability" of features, is necessary if differential survival in one generation is to affect the characteristics of the organisms of the next generation.)

Although Darwin did not do so in any clear way, it will help us to separate two intertwined concepts pertaining to adaptation. The first is the relative engineering adequacy of a design (given the relevant constraints) as a solution to a particular problem. The second concerns the process by which the design was produced. On this usage, if the variations of a given feature, system, or behavior pattern were causally efficacious in the refinement of that feature, system, or behavior pattern by means of natural selection, then that feature counts as an adaptation relative to its alternatives. However, it will not do to say simply "if a feature was produced by natural selection, it is an adaptation"; account must be taken of the "mysterious laws of correlation of growth." Darwin rightly emphasized this point; so far as the process of natural selection is concerned, many features are epiphenomenal, or are produced despite being mildly deleterious, or arise independently of their slight usefulness, simply because their selective consequences are far smaller than those of correlated features. Given what we now know about pleiotropy, multiple functions of organic structures, multigenic determination of features, and so on, we should recognize the broad application of this point. In other words, one is making a substantial claim if one main point is that a feature is an adaptation in the *process* sense. One is claiming not only that the feature was brought

[6] For sufficiently long-lived organisms, the *range* of problems faced should not change greatly from generation to generation. But for short-lived organisms, the environment will vary, both randomly and cyclically, on a scale of many generations; the variation will be life-threatening but is likely to fall within roughly predictable limits. Indeed, typical cyclic variations (e.g., seasonal changes) are associated with cues (e.g., changes in the number of daylight hours) that are utilized by short-lived organisms (e.g., as a signal to produce an overwintering form). Although the environment has many components that are random *with respect to the organism*, other things being equal, those organisms with design features that enable them to utilize or withstand the fairly regular excursions of the environment are better adapted – and more likely to leave descendants – than those that are not.

about by differential reproduction among alternative forms, but also that the relative production of the feature vis-à-vis its alternatives played a significant causal role in its production. This claim is a historical claim, subject to all the epistemic difficulties attendant on such historical claims (Gould and Vrba 1982, p. 182).

Incidentally, this discussion reveals two more touchstones for analyzing adaptation. The first is that an analysis of the process of forging adaptations must make it possible to distinguish selective processes from random processes, design from drift, and effects of natural selection from effects of mere *de facto* differential reproduction. Darwin's way of achieving this is, effectively, to distinguish between differential reproduction and natural selection. For Darwin, natural selection is systematic differential reproduction due to the superior engineering fitness of certain available variants – due, that is, to the relatively better design of the favored organisms.

The second touchstone cannot be stated comfortably in Darwinian terminology. It is simply this: Claims about relative adaptation concern, in the first instance, a relation between a phenotype (or range of phenotypes) and the environment relative to a field of alternative phenotypes. Thus, although genes, gene complexes, genotypes, and so on may be said to be well or ill adapted and may be compared with respect to their degree of adaptation, within *Darwin's* theory, such claims ought to be treated as depending on analyses of the various advantages manifested by the (range of) phenotypes that the relevant genes or genomes produce in the appropriate (and appropriately weighted) environmental circumstances. This is true whether one speaks statically of the degree of adaptedness (or co-adaptedness) of a gene or gene complex or dynamically of the process by which genetic adaptation is achieved. *Full analysis of adaptation cannot bypass the phenotype.* To be sure, the phenotype alone will not suffice; one must decompose phenotypic variation into heritable and nonheritable components. The former is the raw material on which Darwin's process of natural selection acts in creating adaptation.[7] Some subsequent concepts of fitness (as we will see later) are used in ways that bypass the phenotype; I argue that they are weaker than the concept we are articulating currently and that the move of bypassing the phenotype raises difficulties that the present concept does not face.

It is time to codify our results concerning Darwin's concepts of adaptation. (For ease of reference, a short version of this codification is given in Table 1.) So far, I have discussed three interrelated concepts. The first might be labeled "limited perfect adaptedness" or "absolute engineering fitness." A (type of)

[7] I thank P. Richerson for this last point.

**Table 1.** ***Concepts of Adaptation***

| Label | Definition | Historical Knowledge Required? |
|---|---|---|
| 1. Absolute engineering fitness or "limited perfect adaptedness" | A type of feature (or organism) manifests an optimal engineering solution (within real design constraints) to a real environmental challenge. | Only to specify the design constraints and the design problems. |
| 2. Relative engineering fitness | The type of feature (or organism) manifests a better engineering solution (within real design constraints) to a real environmental challenge than specified alternatives. | To specify design constraints, design problems, and the range of alternative types. |
| 3. Selected engineering fitness | The characteristics of the type of feature (or organism) are present as a consequence of their higher engineering fitness with respect to real environmental problems (faced in the evolutionary history of the organism) as compared with the historically available alternative types or characteristics. | To specify design constraints, design problems, the range of alternative types, and the causal history behind the prevalence or fixation of relevant characteristics. |
| 4. Realized fitness ("Darwinian fitness," "tautological fitness") | An organism (or class of organisms sharing some property) has higher realized fitness in environment E than alternative organisms (or classes of organisms) if and only if its actual rate of reproductive success is higher than those of the alternatives. | Only to specify actual rates of reproductive success. |
| 5. Expected fitness ("Darwinian fitness") (propensity for greater reproductive success *de facto* rather than by virtue of design considerations) | A type of organism (or other replicating entity) has higher expected fitness than its competitors in environment E if and only if it has an objective propensity to out-reproduce them in E. Usual measure: relative reproductive success within replicate populations in "identical" environments. | Only to specify actual rates of reproductive success of replicate populations and to certify comparability of environments. |

feature or a (type of) individual possesses absolute engineering fitness if and only if its design manifests an optimal engineering solution to the appropriate (real) challenge or range of challenges posed by the environment. Following Ospovat, I suggested that in developing the theory of the *Origin*, Darwin replaced this concept with one which might be appropriately labeled "relative engineering adaptedness" (or fitness). A (type of) feature or a (type of) individual possesses higher relative engineering fitness than an alternative type if and only if its design manifests a better engineering solution within the appropriate (real) design constraints to a specific (real) challenge or range of challenges posed by the environment. Finally, a (type of) feature is an adaptation if and only if its design characteristics were produced as a causal consequence of their relative engineering fitness as compared with those of the relevant alternative types, as a solution to a problem or range of problems posed by the environment in the evolutionary history of the organisms in question. One may be able to explain the fact that a (type of) individual is relatively better adapted than alternative (types of) individuals if and only if its greater relative engineering fitness is a consequence of specific adaptations. This yields a third sense of fitness, which I call (relative) "selected engineering fitness."[8]

Again, it is important to realize that traits with high engineering fitness need not have been produced by natural selection. The fact that a trait confers advantage with respect to survival or reproductive success does not by itself justify any claims about its historical origins. The notorious opportunism of evolution amounts to an ability to turn traits to advantage no matter how they originated, no matter how they were initially produced (see Williams 1966, p. 12). Accordingly, it is crucial to understand that the theory of the *Origin* requires one to connect the concept of adaptation with that of natural selection. The mere fact that variant forms are differentially fit – that is, that they exhibit differences in relative engineering fitness – is not enough to justify employment of the concept of selected engineering fitness in evolutionary theory.[9]

[8] It is this notion of adaptation that Williams has in mind when he says that "evolutionary adaptation is a special and onerous concept that should not be used unnecessarily, and an effect [i.e., a fitness-increasing use to which a trait is put] should not be called a function [a *designed* fitness-increasing use] unless it is clearly produced by design and not by chance" (Williams 1966, p. vii).

[9] Contrast Gould and Vrba (1982), who argue that many traits have been "coopted"; that is, that their current uses are "effects" not "functions." To work out the problems for evolutionary theory raised by such a position, it is necessary to distinguish between traits that were originally shaped by natural selection for their current use and those that originated as correlated characters or via selection for a *different* use. Gould and Vrba wish to restrict the trait-descriptor "adaptation" to *traits shaped ab initio for their current use by natural selection* and to employ the neologism "exaptation" for traits coopted for new uses (see their Table 1). I do not like this terminology,

This becomes clear when one realizes that Darwin's earlier theory used differences in relative adaptedness (i.e., engineering fitness) as the means of restoring (limited) *perfect* adaptedness – that is, absolute engineering fitness – but derived the direction of evolution from the direction of variation. In contrast, the central concept of Darwin's theory in the *Origin* is the concept of features of organisms with high relative engineering fitness that have been designed by the historical processes involved in natural selection rather than by directed variation. This is what later writers like Ernst Mayr mean by the creative power of selection; this is "the onerous concept of design" that G. C. Williams seeks to restore to prominence in evolutionary biology. This is the feature of Darwin's theory that his contemporaries found least persuasive.

## CONCEPTS OF ADAPTATION IN NEO-DARWINISM

I turn now to the so-called genetical – or synthetic or neo-Darwinian – theory of evolution by natural selection articulated in the 1930s and 1940s and the regnant orthodoxy since the 1950s. My strategy is to exhibit two further interconnected concepts of fitness, demonstrate *that* and *how* they are interconnected, and show that failure to recognize the complexity of their interactions with each other and with the Darwinian concepts pertaining to adaptation results in vulgarization of evolutionary theory. On the present occasion, I proceed somewhat dogmatically, with only slight documentation of the use of the alternative concepts to be discussed.

There are many versions of the synthetic theory and many accounts of its central features.[10] For present purposes, one may venture to say that it placed the Darwinian theory on a new foundation – to wit, on a foundation of Mendelian population genetics. A significant dispute, though not central here, concerns the relative importance of population genetics versus that of the principle of natural selection. Ruse, for one, sees natural selection as a *consequence* of the principle of population genetics, whereas Mary Williams – and I – see population genetics as the contingent elaboration of the consequences

but their claim that we need some such distinction and that we should mark it with appropriate terminology is well taken. Indeed, such a distinction is of critical importance if we are to evaluate the Darwinian claim that adapted features in a static sense (characters with high relative design adequacy) are largely adaptations in an evolutionary sense (traits shaped by natural selection *in the first instance* to perform those tasks that they now perform).

[10] The best single source (though its coverage is partial and biased) is Mayr and Provine (1980). Examples of the controversies over the precise content of the synthetic theory are Gould (1980, 1981); and Orzack (1980).

of natural selection for a limited class of organisms (Ruse 1973, especially pp. 48 ff.; Williams 1973, especially pp. 86–8).[11]

Whatever the issue of that dispute, however, one consequence of the central role of population genetics in the synthetic theory has been a series of shifts in the concepts of fitness, adaptedness, and adaptation and in the measures of fitness and adaptation employed in evolutionary investigations. These shifts are complicated and have resulted in considerable confusion on the part of both biologists and those analyzing evolutionary theory. Here, I attempt to disentangle only a few of the central confusions.

Both the following senses of "fitness" have often been labeled "Darwinian fitness" in the recent literature. The first of the two applies in the first instance to individual organisms. According to this sense of fitness, an individual (or a class of individuals) is more fit (or better adapted) than its competitors, not if it has a higher expectation of survival and reproduction in virtue of its design, but simply if it, in fact, enjoys relatively greater reproductive success. Degree of adaptation, in this sense, is an empirical property of the organism (or of the class) in question but one that can only be known *post hoc*.

It is ironic that this concept has come to be labeled "Darwinian fitness." Darwin almost certainly meant the phrase "survival of the fittest" to stand for the tendency of organisms that are better engineered to be reproductively successful. Because the use of the present concept turns the phrase "survival of the fittest" into the tautology "the reproductively successful are reproductively successful," one traditional label for this kind of so-called Darwinian fitness is "tautological fitness." I prefer the less tendentious label "realized fitness," especially because *realized fitness is an empirical property* – namely, *relative reproductive success*.

It should be clear that this sense of fitness falls victim to Scriven's thought experiment concerning the pair of twins. Although relative (actual) reproductive success is, I repeat, an empirical property of the relevant individuals, it simply does not measure degree of fitness or adaptedness in any of the senses relevant to evolutionary theory – all of which have something to do with *systematic* or *designed* fit with the environment, with causally mediated propensities to reproductive success.[12]

[11] Ruse has reaffirmed the stance he took in 1973 in a number of recent writings. Williams's views on this topic are never stated as directly as Ruse's. My interpretation of her position rests on a reading of a number of her other works and on the general character of her views in addition to the passage cited. Incidentally, despite quite a different starting point and style of argumentation, her interpretation of the confusions surrounding the use of the concept of fitness in the synthetic theory (Williams 1973, pp. 88–100) is quite close to the one put forward in this chapter.

[12] P. Richerson suggests that realized fitness (i.e., relative role of actual reproductive success) applies only to organisms exhibiting *heritable* differences. If the concept is so restricted, the

There is, however, a much more interesting concept, also sometimes called "Darwinian fitness," that plays a prominent role in the literature. I am grateful to Francisco Ayala for forcing me to recognize its importance. Although it is frequently used, it has only recently been articulated with some clarity in the literature (Brandon 1978; Mills and Beatty 1979),[13] so I discuss it with care.

It is possible to recognize and even to work up a good quantitative estimate of the propensity of a type of organism to out-reproduce its competitors in a certain environment or range of environments even though one can offer no historical, causal, or design-based analysis of that propensity. Consider an idealized experiment in which, say, 100 replicate cultures of *Drosophila melanogaster* are begun, each with 50 percent flies of type A and 50 percent flies of type B. If 50 of these cultures are raised at high temperatures and 50 at low temperatures in otherwise standard environments and, after 20 or 100 generations, a roughly stable equilibrium is reached such that at high temperature roughly 85 ± 10 percent of the flies in each culture are of type A and at low temperature roughly 65 ± 10 percent of the flies are of type B, the natural conclusion would be that type A flies are better adapted to (i.e., have greater engineering fitness in) the warm environment, whereas type B flies are better adapted to the cool environment.

Unfortunately, as Guyot and Lewontin have reminded me, life is not so simple. In our hypothetical experiment, type A flies out-reproduce type B flies at the warm temperature *when the ratio of the two types in the population is 50:50*. But, when the population is at equilibrium (85A: 15B), their relative reproductive rates are equal. Therefore, the selection coefficients are frequency-dependent; the constitution of the population alters the likelihood of relative reproductive success for the two types. If type A flies were always likely to out-reproduce type B flies in a stable warm environment, over a long enough period, type A flies would entirely replace type B. This shows how crucial it is to consider the population of conspecific organisms in describing the environment.

For these calculations, it does not matter whether the flies in question are identified by their phenotype or their genotype, although it does matter

present counterexample is avoided, for there is no heritable difference between identical twins. But the point of the example remains unaffected if we replace our twins with full or half siblings – or even unrelated individuals. One-time rates of reproduction, whether of individuals or individuals belonging to a certain group or kind, are subject to chance effects (e.g., lightning bolts and sampling errors) and extraordinary occurrences. For this reason, they provide uncertain evidence of the propensities or tendencies underlying the actual reproductive rates.

[13] But for important criticisms of the concept of fitness described by Brandon and Mills and Beatty and in this essay (criticisms whose fundamental point I reject), see Rosenberg (1982, 1983). I thank Professor Rosenberg far supplying me with advance copies of his papers.

that at least some of the differences between them must be heritable. Either way, in the new sense of fitness, organisms of type A are relatively fitter than organisms of other types in environment E if and only if in E, organisms of type A have an objective propensity to out-reproduce those of the alternate types. Usually, the best experimental measure of such a propensity is a statistically significant difference in the reproductive success of replicate populations.[14] Although this notion is now commonly call "Darwinian fitness," a more useful label found occasionally in the literature is "expected fitness." To repeat, the (relative) expected fitness of a type of organism as compared with specific competitors in a specified environment is its propensity to manifest a certain (relative) rate of reproductive success as compared with those competitors.[15] In a secondary sense, a trait of the organism may be said to make a certain contribution to its fitness according to the change in fitness – that is, the change in the propensity to reproductive success – that is correlated with its presence as opposed to its absence or the presence of some specified alternative(s), as appropriate.

Usually the best experimental measure of such a propensity is a statistically significant difference in the reproductive success of the relevant types of organism in replicate populations. In a good experimental study, the replicability of the results allows a powerful statistical argument to the effect that there is an objective tendency at stake rather than a random or unique phenomenon. Thus, in the case discussed, in the warm environment with 50 percent of both types of flies present, type A flies have greater expected fitness than type B flies; with 85 percent of type A flies present, both types have equal fitness. Such an argument removes the central objection to realized fitness: If we use the various track records of replicate populations in a particular environment as a measure of the relative fitness of various types of organism in that environment, it quickly becomes obvious that in any *single* run, organisms of the relatively fittest type may not out-reproduce their competitors; indeed, there are occasional cases in which none of the fittest organisms survive. Thus,

[14] There are serious technical difficulties with such measures. Important discussions of some of these may be found in Prout (1969, 1971a, 1971b) and Lewontin (1974). Many of the problems covered by these authors are particularly acute when one deals with natural populations. Some of them can be circumvented in properly designed laboratory experiments – at the price of relativizing their direct testimony about fitness to laboratory but not to natural environments. The difficulty, often serious, remains that traits of considerable importance in laboratory populations and environments may not be of comparable importance in natural populations and environments.

[15] Actually, for technical reasons discussed in Mills and Beatty (1979, pp. 247 ff.), it is preferable to employ a *distribution* of propensities to leave various numbers of offspring. For present purposes, however, we gain considerable clarity by abstracting from these complexities, which do not affect the central points being made herein.

the claim that the fittest did or will survive in any particular run, even one lasting hundreds of generations, is not a tautology. Nor is it a tautology that the survivors in any natural population are or were the fittest.

Careful attention to the distinction between realized and expected fitness is crucial to a proper understanding of the literature of the synthetic theory in at least two ways. The first of these is the elimination of serious conceptual confusions from the classic texts. The confusions that result from conflating the two senses of fitness just introduced have given needless support to those who charge the synthetic theory with mistaking tautologies for empirical claims (and worse). I do not document these confusions at length (although innumerable examples can be found in the dreary literature affirming – or denying – that evolutionary theory is tautological), but I do illustrate the matter by reference to two passages written by founding architects of the synthetic theory.

In Dobzhansky (1970), there is a discussion of the various differences among members of a population that "influence the contribution that the carriers of a given genotype make to the gene pool. This contribution [i.e., the *actual* contribution], relative to the contribution of other genotypes in the same population, is a measure of the Darwinian fitness of a given genotype" (p. 101). In the terminology introduced here, Dobzhansky seems to be claiming that realized fitness is a measure of expected fitness. In a somewhat stronger passage, Simpson (1949) claims the following:

> It must, however, be noted that the modern concept of natural selection . . . is not quite the same as Darwin's. He recognized the fact that natural selection operated by differential reproduction, but he did not equate the two. In the modern theory natural selection is differential reproduction plus the complex interplay in such reproduction of heredity, genetic variation, and all the other factors that affect selection and determine its results (p. 268).[16]

A full reading of the books from which these quotations are taken, as well as of other major works by their authors, reveals that despite verbal confusions, *in fact* (most of the time) they employ the concept here labeled "expected fitness" rather than that of realized (or tautological) fitness. Thus, pages 219–29 in Simpson (1949) constitutes an extended argument that natural selection is a "systematic" and "orienting" process, that "the correlation between those having more offspring, and therefore really favored by natural selection and those best adapted, or best adapting to change [i.e., those that have high expected fitness] is neither perfect nor invariable, [but] only approximate and

[16] I thank G. Simmons (University of California, Davis) for calling this passage to my attention.

usual" (Simpson 1949, p. 221); that the *statistical* bias of "selection... [as] a process of differential reproduction" (p. 224) favors "well-integrated" organisms (p. 224) and "favorable or adaptive combinations" of genes (p. 225). Similarly, in the third edition of his path-breaking *Genetics and the Origin of Species*, Dobzhansky (1951) defined the Darwinian fitness of a genotype as "the relative *capacity* of the carriers of a given genotype to transmit their genes to the gene pool of the following generations" (p. 78, my emphasis). And, in a passage of his 1970 book preceding the one quoted previously, he had accepted Lerner's definition of selection as "*non-random* differential reproduction of genotypes" (Dobzhansky 1970, p. 97, my emphasis). Indeed, considerable portions of the four chapters on selection (and fitness) in the later book are devoted to illustrating the statistically systematic character of the process and properties in question. The removal of the conceptual and verbal confusions in such passages, though an unexciting task, contributes significantly to an articulation of the synthetic theory able to withstand misguided criticisms.

A rather deeper contribution of the articulation and clean separation of realized from expected fitness is an improved understanding of the difficulty of applying the synthetic theory to natural populations. One piece of this puzzle stems from the peculiar way in which application of the concept of expected fitness plays an intermediate role between application of the concepts of engineering and realized fitness. The interplay among these three concepts is subtle and interesting. On the one hand, expected fitness resembles engineering fitness (relative adaptedness) in designating a *propensity*, shared by the members of a kind. On the other hand, it resembles realized fitness in that it does not entail or involve any kind of design analysis; the propensity in question is simply to have thus and such a relative reproductive rate in the relevant environments and relative to the specified competitors.[17] Like realized fitness, expected fitness is concerned with track record, not design; like engineering

[17] F. Ayala (University of California, Davis) suggested the following example to illustrate that increased expected fitness may, in specifiable circumstances, be associated with *reduced* engineering fitness. Two species (e.g., of beetles) compete in a population cage. One is regularly able to displace and eliminate the other. However, when a certain mutant (manifesting, say, red eyes) appears in the population of the fitter species, it displaces the wild type and then is displaced and eliminated by the erstwhile less-fit competitor. This is precisely what happens in experimental populations of *Tribolium castaneum* and *T. confusum* as briefly described in Ayala and Valentine (1979). The original report is in Dawson (1969). Ayala's (unconfirmed) hypothesis is that the red-eyed mutant emits a systemic poison to which its conspecific competitor is highly sensitive, itself is moderately sensitive, and the competing species is only slightly sensitive. If this hypothesis is correct, the increased expected fitness within the conspecific population of the mutation causing red eyes is associated with a decrease in the engineering fitness (and degree of adaptedness) of its carrier, for the organism has been weakened by the mutation. Thus, increase

fitness, it deals with a propensity (i.e., a propensity to have a certain track record) rather than actual success or failure on a case-by-case basis.

When one deals with natural populations, it is nearly always impossible to replicate the environmental conditions (including presence of and comparable opportunity for interacting with other organisms) and the genetic or phenotypic constitution of the original population. Because replicate cultures or populations in "the same" environment are unavailable, it is extremely hard to obtain adequate measures of propensities to reproductive success from field studies. Yet, a biologist who knows his or her organisms well is often (rightly or wrongly) confident that the reproductive outcome of a particular case study reflects the design of the organisms (engineering fitness) and will, thus, be tempted to treat the actual reproductive success rates of those organisms (realized fitness) as a reflection or measure of their objective propensity to success (expected fitness). This is probably why Dobzhansky took the *actual* contribution per genotype to the next generations to be a measure of the "Darwinian" fitness of those genotypes.

The resultant conflation of actual differential reproduction (realized fitness) with a systematic propensity for such differential reproduction (expected fitness) poses serious and often unnoticed epistemological difficulties. This becomes apparent when one notices that if one prevented the use of tacit background knowledge and pretended that one knew only the actual survival or reproduction rates, the support for the inference from realized to expected fitnesses would virtually disappear.[18] *Yet, almost nothing is known about the reliability of our beliefs about the extent to which "chance" events affect reproductive success from case to case.* Nor do we have a good way to discriminate interactions in which, effectively, organisms with constant fitnesses have been exposed to relevantly changing environments from interactions in which, effectively, the relevant fitnesses have varied in frequency-dependent ways.

This argument shows that field studies (which are automatically fraught with unique occurrences, non-replicabity of the relevant populations, and limited repeatability of the biotic and other environmental circumstances) provide a rather slender basis for estimation of expected fitnesses. When such

in expected fitness need not – and (I am confident), in real life, at least sometimes does not – correlate with improvement of design.

[18] As the preceding footnote shows, both the inference from known aspects of the relative engineering fitness of a group of organisms to relative reproductive success and the converse inference are quite chancy. When one adds that a broad distribution of outcomes in replicate populations is compatible with fairly extreme differences in Darwinian fitness (especially when the environment is variable or "patchy"), it is clear that inferences between actual reproductive outcome and reproductive propensities require considerable support.

estimates are challenged, about the only way to bolster them is by showing that the actual survival and reproduction rates reflect *known* differences in the design-based abilities of the organisms to overcome the challenges of the environment or their competitors. That is, information about reproductive outcomes is supplemented by information regarding engineering fitnesses; a sufficiently powerful analysis of *engineering* fitness supports the use of actual reproductive outcome as an estimate of *expected* fitness. (So, too, does a consistent pattern over several generations, given adequate environmental stability.) Biologists have, of course, often argued in support of fitness estimates along these lines, but only a handful of cases have been worked out in adequate detail to be generally acknowledged as fully persuasive. Perhaps the best of these stems from the studies of industrial melanism of the moth *Biston betularia* (and some other species), conducted largely by H. B. D. Kettlewell and his colleagues and described in virtually all modern textbooks of evolution (Kettlewell 1961, 1973). The repetitious centrality of the example of *B. betularia* in the standard texts provides a crude and informal measure of the difficulty of demonstrating *in convincing detail* that the actual changes in the composition of natural populations reflect propensities (expected fitnesses) or are truly consequences of the design features (engineering fitnesses) of the organisms in question. And, it is of rather considerable interest that in the case of *B. betularia* (and in many other interesting but less fully elaborated cases), the primary support for the inference to expected fitnesses consists of showing that the actual outcomes conform to the expectations generated by an analysis of engineering fitnesses.

This analysis of some of the conceptual pitfalls encumbering the application of contemporary concepts of adaptation enables us to articulate and support one of the major concerns in Williams (1966). Williams sought to restore the importance of design considerations in evolutionary theory, emphasizing the importance of "historical" analysis of the problems faced by the relevant populations in their evolutionary past. It requires a (partly) "historical" analysis of the problems faced by organisms that belonged to the relevant populations to forge the connections between current design (evolved function) and the process and outcome of natural selection. In my idealized *Drosophila* experiment, in the case in which type A flies completely supplanted type B flies in warm environments and vice versa in cold environments, one can draw the conclusion that flies of type A are better adapted to the cold environment. Yet, whereas it is almost certainly true in such a case that (if one holds other environmental factors constant) flies of type A have relatively higher expected and engineering fitnesses in the warm environment, it in no way follows that the recognized distinctive features of type A flies constitute an adaptation to warmth and the distinctive features of type B flies, an adaptation to cold. For example, the problems faced by each type of fly may revolve around temperature-sensitive disturbances of reproductive physiology that are only accidentally correlated with our ability to differentiate type As from type Bs or on some other more or less bizarre possibility. Such issues can be resolved (if at all) only by a (difficult) study of the evolutionary history of each organism. Typically, thanks to our ignorance, we substitute plausible scenarios (what Gould and Lewontin, 1979 call "just-so stories") for such history. Needless to say, the evidential worth of such scenarios is subject to challenge.

In the absence of firm knowledge of the evolutionary history of our flies and in light of the multiple selective effects that *may* have operated on them in natural as opposed to laboratory environments, it remains an open question whether the greater expected fitness (and supposedly greater engineering fitness) of the type A flies in warm environments was shaped by differential reproduction operating on the favorable response of the ancestors of the type A flies to warmth (in which case, they are exhibiting selected engineering fitness); whether what is involved is really frequency-dependent selection; whether their advantage is an example of "the mysterious laws of correlation" together with the operation of selection on *other* consequences of the type A constitution; or whether their advantage is a consequence of some sort of random change – for example, a series of point mutations that, taken singly, were selectively neutral *in the natural environments of the flies*, although they are highly consequential in the chosen laboratory environments.[19] To repeat Williams's formulation from the précis of his book: "Evolutionary adaptation is a special and onerous concept that should not be used unnecessarily, and an effect should not be called a function unless it is clearly produced by design" (Williams 1966, p. vii).

## "FITNESS," "ADAPTATION," AND RECENT EVOLUTIONARY CONTROVERSY

Having completed my direct discussion of the principal uses of the terms "adaptation," "adaptedness," and "fitness" in evolutionary theory, I devote the final section of this chapter to an exploration of certain ramifications

[19] This discussion of the *Drosophila* experiment is easily summarized in Gould and Vrba's terminology: the outcome of the experiment leaves open the question whether the greater expected fitness (and the putatively greater engineering fitness) of type A flies to a warm environment is an adaptation or an exaptation.

of my analysis and of the value of keeping the various concepts of fitness properly separated. I make four major points.

First, in light of my discussion, it is easy to see that many recent controversies over the status of evolutionary theory turn on a vulgarization of that theory. As soon as evolutionists employ the concept of "realized fitness" in lieu of the concept of an objective propensity toward reproductive success, they lay themselves open to the charge that their theory has no empirical content. As soon as natural selection is characterized as differential reproduction rather than as differential reproduction *in consequence of systematic or design differences*, one is barred, by one's mistaken choice of definitions, from distinguishing between selective and random or accidental cases of differential reproduction. The subtle difference between using realized fitness as a contingent measure of expected (or engineering) fitness and of replacing the latter concepts by the former is, therefore, of considerable importance. One must be on guard against confusing the two moves.

The use of actual reproductive outcomes as rough measures of (but not as substitutes for) expected or engineering fitnesses in the study of natural populations can, indeed, be justified in appropriate circumstances. What biologists must recognize is that *in order to justify the use of de facto reproductive success in actual cases as a measure of evolutionary fitnesses, they must provide good grounds for believing that (in the instances in question) the actual reproductive successes are manifestations of the relevant propensities.* This requirement is comparable to the need to prove an existence and uniqueness theorem when one defines rational numbers in terms of sets of ordered pairs of integers;[20] the procedure is legitimate when the justificatory requirement is met, otherwise not. Conventional wisdom in evolutionary theory holds that that condition is often met – that the organisms within an undisturbed population are usually at, or near, a local maximum of expected fitness and are unaffected by unusual or stochastic factors. I do not know (and it is not easy to establish) how often such claims are correct, but when they are, the measurement of approximate expected fitnesses by means of tautological fitness is a sound procedure. Critics of this procedure should recognize that, in the appropriate circumstances, *this procedure does not bypass the use of both of the concepts of expected and engineering fitness.*

[20] Thus, one most show that there is a number defined by each and every properly constructed set of ordered pairs (e.g., in some systems, {(1,2), (2,4), (3,6)...} defines the number ½) and that each and every such set picks out a unique number. Once this is proved, symbols such as "½" may be used in an unambiguous way in the system in question. Similarly, once it is shown that the actual reproductive rate probably reflects the relevant propensities, that rate may be used as an approximate measure of those propensities.

This discussion makes apparent both how easy it is to fall into conceptual confusion here and that the confusion (despite its prevalence in the literature) is avoidable. It is the concepts of expected and engineering fitness of which evolutionary theory makes essential use, and a proper analysis of the structure of the theory should address the role of these concepts rather than that of realized fitness. In the remainder of my discussion, I proceed accordingly.

Second, in general, it is as difficult to justify the step from expected to engineering fitness as it is to justify the step from realized to expected fitness. This is especially apparent when one considers discussions of fitness at the level of genes and gene complexes. Although we know which alleles (or variants of a gene complex) have the highest expected fitness in a given environment, we often do not know the precise reasons for these relative fitnesses. Because many (probably most) genes have pleiotropic effects when they are expressed, because their precise expression (if any) depends on the presence or absence of other genes, and because the contribution to fitness of their expressed effects may well vary at different stages of the life cycle, the inference that a particular effect is the gene's primary contribution to fitness is quite substantial.[21] Thus, the concept of expected fitness discussed in this chapter is considerably weaker (and more general) than Darwin's concept of engineering fitness. Accordingly, because the synthetic theory uses the former concept, its doctrine regarding adaptation is somewhat weaker than Darwin's. On the synthetic theory, the forces involved in natural selection stem from differences in expected (rather than engineering) finesses. Because differences in engineering fitness constitute *ipso facto* differences in expected fitness but not conversely (see note 17), in principle the synthetic theory allows a broader range of causal antecedents to yield adaptations. Whether the weaker forces permitted by the synthetic theory, primarily at the genic level, in fact play an important role in the production of adaptations is not an easy question. To my knowledge, it has not been directly addressed in the literature.

[21] There are exceptions, of course – for example, when the effect is a dramatic reduction of viability or loss of function. But, even then, matters can be complicated – for example, the familiar case of the gene causing sickle-cell anemia whose positive contribution to fitness when in heterozygous condition was not at all easy to locate. And, notice that it is really the *genotype* (homozygous normal, heterozygous normal and sickling, homozygous sickling), not the gene, that is *causally* relevant to the fitness of the carrier. This raises problems (several of which will be mentioned) treated usefully by Wimsatt (1982, especially Sections 2 and 3), and elegantly developed by Sober and Lewontin (1982). Unfortunately, I read these papers too late to shape the text of the present section. (I thank Professors Wimsatt and Sober for sending me their respective preprints.)

At this point, it is useful to review what is involved in claiming that a trait is an adaptation. There are two components to this claim:

1. The trait in question is an optimal – or, at least, the relatively best – engineering solution of a real problem bearing on (a) survival of the organism; (b) incorporation of energy or other environmental resources into the organism; or (c) incorporation of energy and other resources from the organism and the environment into viable offspring. *All* organisms face problems in all three of these categories, and those that have relatively better (or even optimal) solutions to these problems are those with greater (or optimal) engineering fitness.
2. The design features that yield the high engineering fitness of the trait have been produced by natural selection – that is, in the synthetic theory, by differential survival of organisms with high expected fitness.

One important strand of biological criticism of the synthetic theory turns on a denial of the easy assumption that the distinctive traits of an organism with high expected fitness are *ipso facto* likely to be (or to yield) adaptations in the technical sense just described. This is not a matter that can be resolved by conceptual analysis; its resolution requires extensive and difficult field and laboratory studies. In particular, when expected fitnesses are ascribed to genes, genotypes, or phenotypic markers whose function (if any) is unknown, the inference to engineering fitness is fraught with peril.[22] There is a gap between the genuine expected fitness of such elements and the proper units of analysis in the treatment of adaptation. *The latter are determined by engineering and causal considerations*. It is this gap that plagues the derivation of engineering from expected fitness; it is only when the units of analysis are at the right level – that is, when the causally relevant units are considered – that the two properties are tightly connected (Sober and Lewontin 1982; Wimsatt 1982).

This brings me to the third major point of this section. It is, indeed, the practice of many biologists to casually assume that any distinctive feature of an organism associated with a reproductive advantage (i.e., high expected

[22] This point has both biological and philosophical content. An example of the former is provided by Thornhill (1979, p. 365): "[The] weakest aspect [of the study under review] is its failure to separate the concept of evolved function from the notion of advantage or benefit [i.e., contribution to expected or engineering fitness] . . . The function of a trait is defined as the advantage that characterizes the selective history of the trait; function tells how the trait has contributed to most effective reproduction over evolutionary time. (The elucidation of function is difficult; see Williams [1966].) *Because function and advantage are used synonymously throughout the book, the authors occasionally (and apparently unknowingly) offer interpretations that imply effective group selection*" (my emphasis).

fitness) for its bearers is an adaptation in the engineering sense. Once one realizes how strong the implied claim is, one recognizes that it deserves serious examination. A development of this theme should shed some light on the recent, many-faceted dispute over the proportion and importance of prominent and biologically interesting features of organisms that are not (in our onerous sense) adaptations. In a controversial paper, Gould and Lewontin (1979) attack the "adaptationist program," arguing that it is overly optimistic to suppose that virtually any prominent or biologically significant feature of an organism is an adaptation – that is, not only that it serves a function but also that it was brought about by the action of natural selection operating on variants of that feature. Many subsidiary controversies, in fact, turn on whether particular biologically significant features are adaptations in this sense. At the genetic level, there is the dispute about "selfish" DNA (i.e., DNA that is of no value to any organism but which hitches a ride on the reproductive machinery of organisms) and the continuing controversy about selectively neutral mutations in genes coding for enzymes. (It may yet turn out that selectively neutral point mutations are required to explain the clockwork regularity of change in DNA sequence versus the stop-and-start irregularity of rates of morphological change, speciation, and so on.) At the level of morphological structure and programmed habit, there are innumerable detailed controversies about the correctness, epistemic status of, and support required for selectionist "stories" about the origin of particular features.

There are also general concerns about the paleontological evidence tending to show that whole suites of features change relatively suddenly and relatively rapidly in correlated ways; it is not clear whether such changes occur, in general, via mutual co-adjustment, feature by feature, or via systematic developmental change. Of particular importance is the question whether, on a paleontological time scale, macroevolutionary mechanisms supplement natural selection in a way that must enter into a correct account of the presence of (certain) major traits in surviving populations. (See Vrba [1980] for a useful review of some of the controversial mechanisms under investigation.) This latter worry is central to the debates about whether one can account for macroevolution by extrapolating what we know about microevolution, a topic that I hope to address in a companion paper. Also of concern in the debate over macroevolution is the effectiveness of group and species selection – that is, selection operating at higher levels than classical natural selection. The resolution of such debates, as should be obvious, requires far more than a conceptually adequate account of adaptation but, in all these cases, resolution of the controversy has been impeded by the sort of conceptual confusions about adaptation that I have been illustrating.

My fourth and final point deals with problems regarding the units of selection. Classical natural selection operates, however effectively or ineffectively, one generation at a time by means of a systematic statistical bias in the reproductive rate of phenotypically distinct individuals. Yet, one finds a great variety of claims in the literature to the effect that units such as genes, gene complexes, cellular organelles, and so on, on the one hand, and groups, populations, and species on the other, are units of selection. Herein, I deal briefly with only one of these.

Like many others, Williams is inclined to genic reductionism (Williams 1966, and elsewhere). He holds that virtually everything one wishes to say about natural selection can be said by treating the single gene as the unit of selection. His definition of a gene is somewhat unusual. It includes anything in the genome "which segregates and recombines with appreciable frequency" (Williams 1966, p. 24) and, therefore, includes, for example, whole inversion loops in the chromosomes of certain *Drosphilae* as genes. (This definitional sleight of hand makes no difference to the fundamental issue.) Specifically, he holds that the central processes of natural selection can all be represented in terms of the mean selective effects of each gene, taken singly. Williams grants that a given allele has different effects in different (conspecific) organisms because of both genetic differences in the organisms and the complexities of epigenesis. (These complexities mean that environmental and developmental differences may enable different copies of the same allele, even when it is in the same genetic context, to have different effects.) Nonetheless, Williams argues, each allele has an *average* phenotypic effect that will determine the net strength of the selective processes acting on it. So far as the genetical theory of natural selection is concerned, the heart of the matter is the *net selective value* of each gene, taken singly.

There are two closely related arguments showing that such extreme genetic reductionism is wrong. Both arguments are based on the claim that in quite common and specifiable circumstances, the mean selective value of a gene, were it available, would not work as a construct for making evolutionary predictions and explanations. The more elegant argument, recently articulated in Wimsatt (1980, 1982), traces back to controversies between Fisher and Wright and between Dobzhansky and Muller (discussed in Chapter 6), though Wimsatt is developing it out of some work by Lewontin. It points out that, in many (arguably quite typical) circumstances, the selective value of an allele per generation is a function of the allelic frequencies at *other* loci. If this is so, and if random or selective pressures affect those other loci independently of the allele being considered, the selective value of that allele will change in each generation as the frequency distribution of alleles at the other loci changes. Any system of bookkeeping that assigns fitnesses or selective values to individual alleles has to take account of allele frequencies *at other loci* and will, thus, only appear to treat each locus independently of the others. In Wimsatt's terminology, even though the mean selective value of the allele may be determinate in every generation, it is a "local," not a "global," value in the genetic state space and, therefore, *extrapolations based on that selective value will have no predictive or explanatory value.*[23] In my terminology, the net selective value of a gene in a given generation, like realized fitness, is a *description* of the reproductive outcome *in that generation only*, not an objective propensity or an enduring property with explanatory value.

The second argument is based on a slightly more direct consideration of the way genes affect phenotypes. To the extent that phenotypes are determined by additive interactions among genes, the mean selective value of a gene is determinate. To the extent that genic interactions are nonadditive (and they are not additive in many, mostly ill-understood, cases – including those involving frequency-dependent interactions, changes in the developmental schedule, and so-called regulatory interactions), the mean phenotypic effect of a single gene is unstable and, hence, ill-defined.

As these parallel arguments show, there are deep, substantial, and probably false assumptions that would have to be true if everything we wish to say about natural selection were to be captured, even in principle, in a treatment based on mean phenotypic effects and the associated selective values of genes taken singly. There are, thus, good grounds for holding that the gene is almost certainly not *the* ultimate unit of selection. This result (which may be generalized to cover broad classes of purported reductions) undercuts the idea that we can work our way back up from single genes to locate the "correct" phenotypic categories in terms of which to describe those features of organisms that have been directly shaped by natural selection. In other words, the atomistic study of single genes and their phenotype effects will not, in general, reveal which features of organisms are adaptations arrived at in the course of natural selection.

The general moral of this story for the problem of the units of selection is not that the gene is *not* a unit of selection. (It is.) Rather, the moral is that there

[23] Sober and Lewontin (1982) point out that genic fitnesses can be parameterized so as to take account of the genetic background (and other disturbing factors). They argue that although such devices allow one to calculate the actual change in gene frequencies from appropriate information, the relevant causal factors generating the selective forces are *genotypes*, not genes (and so on in other cases), so that the *explanatory* use of fitnesses *in the relevant cases* (one is mentioned in Note 21) occurs at the genotypic (or appropriate other) level, not the genic level. Wimsatt (1982) makes a similar point.

are units of selection at many hierarchical levels and that we must take into account interactions that cross these levels. The result will be a considerable, but inevitable, increase in the complexity of evolutionary theory, for it means that we will have to examine the strength of the interactions between selective processes at various levels (at least from the level of the single gene to the level of species selection). I, for one, will be much surprised if it turns out (as, of course, it might) that classical natural selection, operating between phenotypically distinct organisms in virtue of their design, is the only type (or the highest level type) of selection we need to consider in reconstructing the history of life. But a full examination of the controversies over the units of selection will have to wait for another day.

## REFERENCES


Ayala, F., and J. W. Valentine. 1979. *Evolving*. Menlo Park, CA: Benjamin/Cummings.

Brandon, R. N. 1978. "Adaptation and evolutionary theory." *Studies in the History and Philosophy of Science* 9: 181–206.

Darwin, C. R. 1859. *On the Origin of Species by Means of Natural Selection or the Preservation of Favoured Races in the Struggle for Life*. London: John Murray.

Dawson, P. S. 1969. "A conflict between Darwinian fitness and population fitness in *Tribolium* 'competition' experiments." *Genetics* 62: 413–19.

Dobzhansky, T. 1951. *Genetics and the Origin of Species*. New York: Columbia University Press.

Dobzhansky, T. 1970. *Genetics of the Evolutionary Process*. New York: Columbia University Press.

Gould, S. J. 1971. "Darwin's retreat." *Science* 172: 677–8.

Gould, S. J. 1980. "Is a new and general theory of evolution emerging?" *Paleobiology* 6: 119–30.

Gould, S. J. 1981. "But not Wright enough: Reply to Orzack." *Paleobiology* 7: 131–4.

Gould, S. J., and R. C. Lewontin. 1979. "The Spandrels of San Marco and the Panglossian Paradigm: A critique of the adaptationist program." *Proceedings of the Royal Society of London, B* 205: 581–98.

Gould, S. J., and E. S. Vrba. 1982. "Exaptation – a missing term in the science of form." *Paleobiology* 8: 4–15.

Kettlewell, H. B. D. 1961. "The phenomenon of industrial melanism in *Lepidoptera*." *Annual Review of Entomology* 6: 245–62.

Kettlewell, H. B. D. 1973. *The Evolution of Melanism*. Oxford: Clarendon Press.

Lewontin, R. C. 1974. *The Genetic Basis of Evolutionary Change*. New York: Columbia University Press.

Mayr, E., and W. B. Provine, eds. 1980. *The Evolutionary Synthesis: Perspectives on the Unification of Biology*. Cambridge: Harvard University Press.

Mills, S., and J. Beatty. 1979. "The propensity interpretation of fitness." *Philosophy of Science* 46: 263–86.

Orzack, S. 1980. "The modern synthesis is partly Wright." *Paleobiology* 7: 128–31.
Ospovat, D. 1981. *The Development of Darwin's Theory*. Cambridge: Cambridge University Press.
Prout, T. 1969. "The estimation of fitnesses from population data." *Genetics* 63: 949–67.
Prout, T. 1971a. "The relation between fitness components and population prediction in *Drosophila*. I. The estimation of fitness components." *Genetics* 68: 127–49.
Prout, T. 1971b. "The relation between fitness components and population prediction in *Drosophila*. II. Population prediction." *Genetics* 68: 151–67.
Rosenberg, A. 1982. "On the propensity interpretation of fitness." *Philosophy of Science* 49: 268–73.
Rosenberg, A. 1983. "Fitness." *Journal of Philosophy* 80: 457–74.
Ruse, M. E. 1973. *The Philosophy of Biology*. London: Hutchinson University Library.
Scriven, M. 1959. "Explanation and prediction in evolutionary theory." *Science* 130: 477–82.
Simpson, G. G. 1949. *The Meaning of Evolution: A Study of the History of Life and Its Significance for Man*. New Haven: Yale University Press.
Sober, E. R., and R. C. Lewontin. 1982. "Artifact, cause, and genic selectionism." *Philosophy of Science* 49: 157–80.
Thornhill, R. 1979. "[Review of] R. W. Matthews and J. W. Matthews, *Insect Behavior*." *Quarterly Review of Biology* 54: 365–6.
Vrba, E. S. 1980. "Evolution, species, and fossils: How does life evolve?" *South African Journal of Science* 76: 61–84.
Williams, G. C. 1966. *Adaptation and Natural Selection*. Princeton: Princeton University Press.
Williams, M. B. 1973. "The logical status of the theory of natural selection and other evolutionary controversies." In *The Methodological Unity of Science*, ed. M. Bunge. Dordrecht, Holland: D. Reidel, 84–102.
Wimsatt, W. C. 1980. "Reductionistic research strategies and their biases in the units of selection controversy." In *Scientific Discovery: Case Studies*, ed. T. Nickles. Dordrecht, Holland: Reidel, 213–59.
Wimsatt, W. C. 1982. "The units of selection and the structure of the multilevel genome." In *PSA 1980, vol. 2*, eds. P. Asquith and R. Giere. East Lansing, MI: Philosophy of Science Association, 122–83.

# 5

# The Influence of the Evolutionary Paradigm[1]

The central concern of this chapter is the relationship between the synthetic theory of evolution and neighboring sciences. In writing about "the evolutionary paradigm" (a title suggested by Max Hecht), I hope to play the role of a philosophical gadfly. In particular, I argue that, in one of the most important ways in which philosophers and scientists use the term "paradigm" these days, *there is no evolutionary paradigm* – and, hence, no influence of "the" evolutionary paradigm.

Obviously, I do not mean to suggest that Darwinism has been without influence. Rather, I wish to probe the way we think about the practice and the unity of evolutionary theory and evolutionary biology in general. I suggest that the muddy concept of a paradigm, as commonly used these days, is one that we can do quite well without, thank you very much. I suggest that bypassing that concept will help us to focus more clear-headedly on the influence of the currently dominant variant of Darwin's theory: the so-called synthetic theory of evolution.

Given the special interest of this symposium in assessing the directions in which evolutionary biology is headed, I emphasize a characteristic of evolutionary theory that I think is of great importance in thinking about its future: namely, its peculiarly *historical* character. I claim that a full appreciation of the nature of historical theories, historical reasoning, and their role in

[1] Originally published in M. K. Hecht, ed., *Evolutionary Biology at a Crossroads* (New York: Queen's College Press, 1989, pp. 149–66). This volume was the product of one of the series of symposia held in honor of the fiftieth anniversary of Queens College of the City University of New York. I am grateful to Marjorie Grene for her criticisms of a draft of parts of this paper and to the National Endowment for the Humanities for support of research in the history of genetics, including evolutionary genetics. The paper has been improved thanks to Albert Cordero, Peter Manicas, and Stanley Salthe, who commented on the paper at the symposium and by some especially helpful criticisms kindly provided by Ernst Mayr.

evolutionary biology ought to shape much of our thinking about the relation of evolutionary biology to other branches of biology and to the sciences more generally. What I have to say about that topic is not particularly original; a good deal of it has been said quite eloquently by Gould and others. But, it is definitely worth trying to make it stick, for – as I hope to show – there is a moral of deep interest to be drawn from Darwinism, a moral that yields some strong suggestions about the directions open to evolutionary theory as we come once again – as we do in every generation – to a major crossroads.

## ABOUT PARADIGMS

Although I pay some attention to the term "paradigm," my principal concern is not semantic but rather with the social and intellectual structure of evolutionary biology. The point of interest can be put in terms of an evolutionary analogy. A species is composed of many variant individuals, often conveniently grouped into varieties. Similarly, there are many varieties of evolutionary biology and many variant positions within each variety. Just as the evolution of species depends on the underlying variation of the individuals of which it is composed, so the development of evolutionary biology depends on the underlying variation among evolutionary biologists. To this extent, it is as worthwhile to ask after the unity in the diversity of evolutionary biology as it is to ask after the unity in the diversity of biological species. The point of worrying about paradigms is to show that some common views regarding the unity of evolutionary biology are not true. Please bear with me, therefore, while I start by clearing some terminological underbrush.

The use of the term "paradigm" in history and philosophy of science traces back primarily to Kuhn's *Structure of Scientific Revolutions* (1962, 1970b). There is a large literature discussing Kuhn's notion of a paradigm. It has been pointed out repeatedly that the term is multiply ambiguous (Masterman 1970). Kuhn himself now admits this and has introduced some new terminology in an attempt to recapture what he takes to be the two fundamental points he was making about the character of major scientific theories and research programs (cf. Kuhn 1970a, pp. 271 ff., and the fuller treatment in Kuhn 1974). The two points are, in fact, fairly straightforward and seem quite commonsensical if we do not read too much into them. They are as follows:

1. Major research programs are built on exemplary texts that record exemplary experiments and interpretative accomplishments. Thus, Newton's *Principia* and Darwin's *Origin* are what Kuhn now calls "exemplars."

These two books showed, by example, how to do science in a certain style – specifically, how to interpret and apply major new theories. One of the terms derived from Greek for an exemplar of this sort is "paradigm." Part of what Kuhn meant to capture by his use of that term is this: *Science is built on exemplars*. Kuhn correctly notes that there are exemplars of many different types, some at a high level and others at a low level. Newton's and Darwin's books are high-level exemplars (i.e., paradigms); low-level ones are ubiquitous in the teaching of science – for instance, the exemplary problems with which we force our students to cope in a freshman lab or the problems and problem solutions in a standard textbook. Such exemplars are the means by which we teach our students to become scientists; they are the cornerstones in the system of apprenticeship that we employ to handle that delicate job.

2. The practices of the scientists working in a given discipline are typically built on a series of complicated implicit agreements and understandings. These are diverse in nature: They cover the basic assumptions of the discipline; the proper interpretation and application of various mathematical formulae; the reasons for choosing one technique or experimental organism over another; judgments regarding the relative reliability of various techniques, laboratories, and even people; judgments regarding the social as well as the intellectual structuring of the discipline; and so on. Sociological frameworks of this sort, which help to structure the work, content, and education in a given field, Kuhn now calls "disciplinary matrices" (Kuhn 1970a, p. 271; 1974, pp. 468 ff.). His original terminology treated the whole range of implicit agreements involved in such socially forged networks as *paradigms* because, he thought, the agreements came about without explicit articulation from the use of exemplars (i.e., the standard texts, techniques, problems, and problem solutions from which one learned as a student of the discipline).

We are now in a position to ask to what extent there is an evolutionary paradigm. Using Kuhn's own separation of paradigms into exemplars and disciplinary matrices, this question is transformed into two questions: Is the practice of evolutionary biology correctly described as based on some central exemplars? The answer to this question is obviously affirmative. One need only mention Dobzhansky, Mayr, Simpson, and Stebbins – or, in a different vein, sickle cell anemia, variation among hemoglobins, and *Biston betularia* – to see how easy it would be to establish this claim. The second question is whether some one disciplinary matrix dominates the practices of evolutionary biologists. This question is a significant one that cannot be properly resolved here. I will, however, offer some good reasons for being skeptical about the presence of anything like a single disciplinary matrix during most of the history of modern evolutionary biology. Not least of these is the fact that evolutionary biology is not a single discipline but rather a complex interdisciplinary field that lies in a region that overlaps onto the territory of diverse disciplines.

## SOME HISTORICAL NOTES

In this section, I offer a rough-and-ready treatment of some historical points that show how implausible it is to treat evolutionary biology before the development of the synthetic theory as falling within a single disciplinary matrix. In the next section, I support the parallel conclusion with respect to the synthetic theory itself.

The large-scale historical point can be put rather simply. Darwin's *On the Origin of Species* (1859) is, without question, an exemplar – it served as a paradigm in the first of the two senses distinguished previously. It is a text to which evolutionary biologists have paid an enormous amount of attention ever since it was published and, to this day, even those evolutionary biologists who have not seriously read the *Origin* are deeply influenced by it. It is a work that has penetrated evolutionary theory and the practice of evolutionary biology very deeply. Nonetheless, it is arguable whether the *Origin* or any other influence has established a dominant disciplinary matrix within evolutionary biology.[2]

Among the many ways of getting at this issue by means of historical investigations, I can pursue only one here – and that only sketchily. This is to examine the relationships among various versions of Darwinism. (It would also be useful to examine the many alternative theories that have played a major role in evolutionary biology at various times and places.) In this connection, it is worth recalling that for a substantial portion of the history of

[2] Ernst Mayr (personal communication) suggests that the *Origin* established a definite theoretical framework or set of principles. He suggests that, after Darwin, descent with modification acquired a central place in biological reasoning that it had not had before and that, therefore, it is a mistake to claim that Darwin failed to establish a dominant disciplinary matrix in evolutionary biology. This criticism is mistaken, for it fails to recognize that a disciplinary matrix (at least, as Kuhn intends the term) is built on common problems, techniques, training, and standards of evaluation established by education *within a reasonably well-defined discipline*. Anatomists, biogeographers, zoologists, botanists, breeders, natural historians, and so on – even if they are evolutionists sharing the *Origin* as an exemplar – do not share a disciplinary matrix. The best reading of the evidence, I believe, does not reveal a common disciplinary matrix for Darwinian evolutionists 1859–1909, even when one restricts one's attention to England.

evolutionary biology, *no* version of Darwin's theory of evolution *by means of natural selection* was the dominant accepted theory. Indeed, in all of the relevant national traditions, Darwinian theory received quite different interpretations so that, looked at from an international perspective, the Darwinian party did not share common assumptions of the sort Kuhn includes in disciplinary matrices. The central point is that it is implausible to claim that there was anything resembling a single disciplinary matrix in evolutionary biology, at least until the general acceptance of the synthetic theory of evolution around the latter 1940s – and, as I argue here, it is not really very plausible then.

In support of these claims, it is useful to begin with Mayr's point that Darwin put forward a congeries of theories (Mayr 1985). For present purposes, I mention only three: namely, that evolution is gradual, that it consists in a process of descent with modification, and that the most significant evolutionary force among the many Darwin recognized is natural selection. I focus on the latter two to some extent in this section. Let us now recall the reception of Darwin's theories in general terms. The initial success of the *Origin* did not (as is sometimes naively assumed) yield widespread acceptance of the theory of evolution *by means of natural selection*; at most, it established that descent with modification could account for the major phenomena of biogeography, paleontology, and systematics, and for the adaptations of plants and animals. To be sure, the theory of natural selection played a major role in this initial success, for it offered a plausible mechanism by means of which evolutionary change could be seen as creating adaptations.[3] Nonetheless, for a long time, the particular mechanisms of evolution were treated extremely pluralistically by most theorists – including Darwin himself, especially in the later editions of the *Origin* and in the *Descent of Man* – and there was an immense amount

[3] Mayr (personal communication) considers this a myth, pointing out correctly that common descent was broadly accepted despite the fact that natural selection was rejected by most naturalists, including even T. H. Huxley. I think, nonetheless, that natural selection "broke the ice" for descent with modification. It made it seem plausible that a thoroughly naturalistic account could be given of adaptations, even to those who rejected Darwin's account of the role of natural selection in evolution. As Lewontin (1983) points out (in an article that Mayr reminded me of!), natural selection broke with the spirit of all prior evolutionary theories (e.g., Lamarck's). Those theories were "transformational" – that is, they required that the evolution within a lineage be a sum of the changes that occurred to individual organisms. Typically, such theories have difficulty with adaptation, requiring ad hoc or secondary causes to explain adaptations. Darwin's "variational" theory, in contrast, took variation as somehow given, so that the source of evolutionary change was some sort of external action on available variation rather than a building up of new permanent variation within the organism for exploitation. (Darwin himself, to be sure, was obstinately pluralist; use and disuse of organs as a source of evolutionary change is a transformational mechanism – quite different, in this respect, than natural selection.)

of debate over the claim that the historical path of evolutionary change was dominantly the effect of natural selection as opposed to saltatory mutation, inherited effects of use and disuse of organs, direct effect of climate, internal orthogenetic drives, and so on.

A central tenet of contemporary Darwinism is that natural selection is the predominant cause of evolutionary pattern. This claim was nowhere fully accepted in Darwin's day; Darwin himself lamented the failure of his work to convince his peers on this score. Indeed, as is shown by an examination of primary sources (Kellogg 1907; Seward 1909) and as is widely recognized in recent historical writings (Bowler 1983; Mayr 1982), belief in the importance of natural selection as a factor in evolution reached a nadir around 1909, the centenary of Darwin's birth and the fiftieth anniversary of the publication of the *Origin*. Although it was nearly universally granted that Darwin's mechanism played *some* role in evolution, by this time there were relatively few defenders of the idea that natural selection is a major contributor – let alone, *the* major contributor – to the fundamental patterns of evolutionary history.

Less well known except among specialist historians is the fact that each of the major national traditions interpreted Darwin quite differently (Glick 1974; Kohn 1985). I discuss, absurdly briefly, England, France, Germany, and Russia to drive this point home. Although I am talking in crude general terms, the sketch offered here is, I believe, an adequate first approximation. In England, descent with modification was rapidly accepted and natural selection taken seriously; the Darwinian theory came to be allied with gradualism and biometry. By the turn of the last century, most Darwinians treated small and copious variations as the raw material for natural selection, emphasized the fine-grained character of adaptation, and stressed the causal importance of the intense Malthusian scrutiny of every variant as the key to evolutionary dynamics. In France, Darwin was treated as a utilitarian cheapener of Lamarck, offering nothing particularly new or interesting outside of the already available evolutionary theories. Even descent with modification had great difficulty in making its way and, insofar as it did, transformational and variational versions of the theory were not properly distinguished (see note 3). A few of Darwin's defenders emphasized his pluralism with regard to the mechanisms underlying evolution, but they conceded that he overstressed natural selection. In Germany, too, the distinction between Lamarckism and Darwinism was not very clear; however, between the influence first of Haeckel and then of Weismann, Darwin's theory became associated with preformationism on two grounds: that the organism contained the history of its lineage (cf. the biogenetic law) and that the developmental unfolding of the organism was a consequence of the inherited materials produced in that history and a necessary condition for the independence of variation from environmental influence.[4]

In Russia, finally, Darwinism was reinterpreted to remove its Malthusian moment. Todes (1987, 1988) reviews this history, arguing that at least two major considerations influenced this stance. One of these is biological – the vast emptiness of the steppes, which were the central arena for the studies of Russian ecologists and naturalists. To the students of the ecology and evolution of the biota of this vast region, struggle against the climate seemed far more intense and important than intraspecific competition. The other influence was, in a sense, political (although common to the political left and the political right) and had to do with the widely acknowledged need to increase population in Russia in order to exploit available resources efficiently. Malthus, it seemed, was just wrong about Russia: Increasing the population under the appropriate social arrangements was the means to guarantee *enough* food for everyone. The main factor limiting the population was insufficient labor to produce food and to develop and maintain the infrastructure to distribute it. Without sufficient population, it was impossible to institute appropriate agricultural and productive techniques, systems, and reforms.

Both of these influences made the Malthusian starting point utterly implausible to Russian theorists and led them to reinterpret Darwin's theory along lines ultimately popularized in the West by Kropotkin – that is, to admit competition with the environment and between species but to insist on the fundamental importance of *cooperation within species*. Such an approach was thoroughly integrated into the strong evolutionary tradition of Russian biology.

Although I have been discussing the differences among national traditions, much the same type of disagreement about the proper content of Darwinism occurred *within* many of these traditions. The differences between transformational and variational theories of evolutionary mechanisms (see note 3) were by no means clear; toward the turn of the last century, almost everyone except Weismann, whose speculations were widely rejected, was confused in this respect. Again, there were serious disagreements regarding just what should be accepted from Darwin and what should count as the core of Darwinian theory. Worse yet, insofar as Darwin's various subtheories were recognized, the interrelations between them became uncertain. Was gradualism required to make sense of natural selection or to support descent with modification?

[4] Mayr (personal communication) correctly cautions that Weismann and other panselectionists were isolated by the turn of the century in Germany. Various forms of orthogenesis and mutationism were preferred by the majority of German biologists with evolutionary interests at that time.

Such questions had become extraordinarily troublesome. Many figures wished to claim the authority of Darwin in support of their own views, while also rejecting – in specific contexts, for specific purposes – various important Darwinian doctrines. The resultant tangles became extremely complicated.[5]

The lay of the land as I am depicting it can be nicely summarized by suggesting that the history of the first 75 or 100 years of Darwinism could be written under the rubric of *Struggling for the Mantle of Darwin*. Throughout this history, the *Origin* and some of Darwin's other writings served as exemplars, but there were constant attempts to reinterpret Darwin's ambiguous and pluralistic theory along widely divergent lines. Thus, the *Origin* was read very differently by different parties, and it was used to support a great many mutually incompatible positions. Frequently, these divergent readings were intended to help forge the sort of consensus that Kuhn characterized as the crucial, although tacit, center of a disciplinary matrix and to convert the resultant consensus into the doctrinal core of evolutionary biology. The fact that many divergent biological disciplines, often drawing on different problems and techniques, were affected by these divergences made the task of reaching full consensus difficult.

Why should *this* exemplary text have served as such a symbol of authority? Because many evolutionists – with the exception of the French and certain of the more extreme Lamarckian and orthogenetic opponents of Darwin – wished to draw on Darwin's reputation as founder of mainstream evolutionary theory and as the most influential naturalist/evolutionist of the age. But, the struggle to cloak such extremely divergent doctrines in the mantle of Darwin, which continues even to this day, is a symptom of the fact that there was and is not a single disciplinary matrix in evolutionary biology. I conclude that the influence of Darwinism on neighboring sciences should not be understood in terms of the influence of "the" evolutionary paradigm.

## THE EVOLUTIONARY SYNTHESIS

The synthetic theory of evolution is a moving target. Nonetheless, in first approximation, it can be characterized fairly briefly, for it was put forward by a series of founding documents in the period from 1937 to roughly 1950. I take the first of these to be Dobzhansky (1937) and the last Stebbins (1950).

[5] For an extreme account of the trouble that results from these considerations, see Hull (1985), who denies that there is no set of shared doctrines adequate to characterize the common features of the different versions of Darwinism.

Among the more important books along the way one must include those by Mayr (1942) and Simpson (1944), though many other writings should be cited as well.

The central point of the synthesis was to demonstrate the adequacy of Mendelian genetics (including especially population genetics) plus an updated version of Darwin's theory of evolution by means of natural selection, joined in the manner illustrated in the founding documents, to serve as the theoretical basis for explaining all evolutionary phenomena. The following doctrines are characteristic of the synthetic theory: the immense variability of natural populations, the genetic (indeed, Mendelian) basis for evolution, the importance of geographic speciation, the adaptive nature of observed differences among organisms, the primacy of natural selection in causing the evolutionary patterns found in the paleontological record, the gradualness of evolution, the compatibility of all macroevolutionary phenomena with the mechanisms of (gradual) microevolution (thus ruling out saltational models), and the importance of what Mayr called "population thinking."[6] Wallace's paper (1989) exemplifies quite well the sort of stance that I would include within the synthesis.

In certain respects, the so-called synthetic theory is better viewed as a treaty than as a theory.[7] By this, I mean that one could not use the population genetic foundations of the theory to predict or retrodict large-scale evolutionary patterns or fundamental features of the taxonomic system. Nor can population genetics alone determine the prevalence, evolutionary importance, or historical trajectory of major traits like sexuality – indeed, it cannot even provide a full answer to Darwin's problem: the origin of species. Nonetheless the synthetic theory did a great deal to reduce the conflict between evolutionists belonging to different disciplines. For example, it served to contravene the paleontologists' belief that Mendelian variation cannot offer a sufficient explanation for macroevolutionary phenomena, to undermine the geneticists' early insistence that genetic (and, hence, evolutionary) change is discontinuous and saltational, and to counteract the systematists' conviction that laboratory experiments on mutations and Mendelian variation were

[6] This list deemphasizes the contribution of genetics to the evolutionary synthesis. It is based on Mayr's (1980) gloss on various contributions to the synthetic theory. Mayr (personal communication) considers the two most important contributions of genetics to the synthesis to be evidence for the facts that "inheritance is particulate, not blending; and that inheritance is hard, not soft (Lamarckian)." He adds that these contributions "have comparatively little to do with population genetics."

[7] The term "treaty" is taken from Depew and Weber (1988). Both that essay and Burian (1988) amplify some of the points made here. This chapter also suggests links to other analyses of the synthetic theory as a metatheory or a schematic theory.

irrelevant to the sorts of variation and evolutionary change found in natural populations. Precisely because the synthetic theory aimed to establish the *compatibility* of the standard population genetic accounts of microevolution with all known evolutionary (especially macroevolutionary) phenomena, it disarmed conflicts between disciplines bearing on evolutionary history. It allowed the claims about the *patterns* and *results* of evolution to be drawn from other disciplines while insisting that the *mechanisms* revealed by population genetics were the only ones needed to bring about those patterns.

Not surprisingly, the evolutionary synthesis has changed with time. For one thing, the tools available to theorists and field workers have grown considerably more sophisticated, especially with the introduction of computers and molecular techniques. One controversial claim of interest to us about the developments from, say, 1950 to 1970 – a claim that I believe is correct – is that the synthesis hardened, at least to some extent, into unexamined dogma, sometimes dismissing viable alternatives on the basis of prejudgment rather than hard evidence (Gould 1983). Early on, the very weakness of the compatibility claim was viewed as one of the *strengths* of the synthesis. In principle, virtually all the known phenomena and patterns that ought to be explained by an evolutionary theory *could* be explained by some variant of the synthetic theory – and so it was not necessary to contemplate or turn to any rival theories. As the synthesis hardened, "could be explained" turned into "are explained," yielding outright dismissal of competing theories.

Although I do not discuss examples in detail, I cite three of the many that could be adequately documented to illustrate the point. These are the treatment by many theorists of the late Sewall Wright's shifting-balance theory (which, after the mid-1950s, was frequently dismissed as placing far too great an emphasis on fixation of random variants in small populations), the change in Simpson's views about quantum evolution and its relationship to natural selection, and the increasing panselectionism of many of the leading figures in the field. (Another symptom of this change, pointed out to me some years ago by Marjorie Grene, is the disappearance from later editions of the founding texts – and from the literature of the theorists of the synthesis generally – of people like Goldschmidt, Robson and Richards, Schindewolf, and Willis. These figures had served as rhetorical targets in the founding debates, and their disappearance signals the fact that their nonselectionist alternatives came to be considered irrelevant and no longer threatening.)

If this characterization of the synthesis as a treaty is approximately correct, it enforces the conclusion that I wish to draw: There is no single disciplinary matrix to which the adherents of the synthesis subscribe. Nor should this be surprising: The scientific advocates of the synthesis are drawn from systematics, population genetics, paleontology, botany, zoology, biogeography, and an immense variety of additional disciplines. To that extent, even though their commitments overlap and even though they share some exemplary texts, their disciplinary allegiances (and their primary training) belong to very different fields. It should not take much reflection on the consequences of this fact to recognize that their socialization as scientists prevents them from sharing a common disciplinary matrix, the supposed disciplinary matrix of evolutionary biology or evolutionary theory. In the last section, I explore a few consequences of this fact for the future of that theory, but first I address one more topic.

## THE SYNTHETIC THEORY AS A THEORY OF HISTORY

The synthetic theory, like Darwinism generally, claims that the details and many of the basic patterns of organismic evolution are, at heart, historically contingent. Gould argued persuasively that Darwin's central accomplishment in this regard was to construct a theory that treated adaptation as a response to the historical sequence of selective demands of the environment (including other organisms), a theory that accounted for taxonomic and morphological order in terms of "historical pathway, pure and simple" (Gould 1986, p. 60). These Darwinian explanations contrast with alternatives couched in terms of "intrinsic purpose and meaning," or of laws of form, and so on. For Darwinians, Gould claims, homology requires and is explained by common descent, whereas similarity of functional form without common descent (analogy), however striking, is accounted for by the adaptive power of selection. Many readers will recognize that this is a substantial and controversial claim, for it commits one to the view that the only correct account of the identity of homologous structures (e.g., of a particular metatarsal in different vertebrates) is identity by descent.

We recognize the occurrence of biological evolution and delineate phylogenies in part by means of "accidents of history." The term is apt. Cladistic classifications, as I understand them, place organisms into taxa defined in terms of shared derived characters ("synapomorphies" in cladistic terminology). Although our assessments of particular traits as synapomorphies may be mistaken, the principle behind such classifications is fundamentally sound – and the corresponding practice seems to me to generate an interesting argument against the derivation of basic evolutionary patterns from the fundamental laws or axioms of evolutionary theory.

Compare, for example, biological with stellar evolution. Stellar evolution is transformational – that is, in the absence of highly unusual interactions with other cosmic bodies, the stages of the life history of each star can be derived from its intrinsic properties (mass and composition) and appropriate initial or boundary conditions (Lewontin 1983, p. 63). In the terminology of some philosophers of biology (Brandon 1982b; 1990, Chapter 3), the effects of such other factors as the *history* of the materials out of which a star was formed (which almost certainly came from other stars) are "screened off" by the physical properties of the star. That is, "accidental" properties like those relevant to evolutionary studies of biological entities are not needed and are of no help in determining either the parentage or the behavior of stars, their likely fate, or the patterns of evolution of populations of stars. In part, this is because stars are sufficiently isolated that the dominant determinants of their behavior, once certain initial conditions have been realized, are fundamental physical laws rather than interactions with their environment or with one another. In part, it is because (in the absence of strong interactions with other stars) the evolution of the ensemble of stars is, in effect, a straightforward summation of the evolution of each star unaffected by interactions.

There are good reasons for supposing that the same is not true for organisms. If the laws of thermodynamics, for example, were powerful enough to determine the patterns of evolutionary history in detail, the evolutionist's use of "accidental" clues would amount to a deep mistake. Instead of supplying crucial information bearing on the behavior and fate of organisms (or other biological entities), it would mask the fundamentally law-driven course of evolution. If laws of form determined ontogenies rigidly, organisms (like stars) would simply have one or another of the available ontogenies, and the transition from one ontogeny to another within a lineage would not be marked by any clues about the history of the lineage. But, organisms and lineages *do* record the accidents of history. The gill slits of mammalian embryos suffice to make the point, although the phenomenon they illustrate is ubiquitous. In the end, both cladistics and evolutionary biology as a whole depend deeply on the contingent fact of evolutionary tinkering (Jacob 1977).

*Because evolutionary theory is concerned, among other things, with analyzing genealogical connections and patterns of genealogical affinity among organisms, it must build on the essential historicity of biological evolution.* The same historicity applies when the aim is to describe patterns of evolution among DNAs, proteins, organisms, taxa, or clades, or to develop evolutionary laws relating, for example, morphological change to cladogenesis or phylogeny.

This historicity is unavoidable; it cannot be escaped by developing a generalized mechanics for evolution. Two examples illustrate the point:

1. The effects of particular molecular mechanisms – and even the content of the genetic code itself – are highly context dependent. Thus, whether a given string of DNA will yield or affect the expression of a particular product depends on the cellular and genetic context within which it is placed. There is no prospect of a generalized mechanics of gene expression powerful enough to take all of the contextually relevant factors into account (except, perhaps, statistically[8]). That is why the analysis of gene expression is a brute-force, messy problem rather than a neat theoretical enterprise.
2. Speciation depends in part on such matters as mate recognition which, in turn, depends on the use of "accidental" characters *by the organisms themselves*. I have in mind H. E. H. Paterson's account of Specific Mate Recognition Systems, and the role it has played in the thinking of people like Elisabeth Vrba and Niles Eldredge.[9] To the extent that separate lineages acquire independent evolutionary fates because the organisms of those lineages employ contingent and accidental differences as cues in mate recognition, the entry of historical accidents into a sound account of evolutionary history is forced on us by the organisms themselves.

This historical component of the Darwinian explanation of underlying form has often been overlooked. Like other historical theories, evolutionary theory must presuppose ahistorical laws as background. These, in turn (assuming they are correct), provide some of the constraints[10] within which history runs

[8] I have in mind results of the sort that Kauffman obtained in simulating some of the general properties of gene-regulation networks. Relatively accessible presentations of some of his results are available in Kauffman 1985 and 1986. Kauffman (1993) brings these lines of work up-to-date. This extraordinarily promising book covers an enormous range of topics – origin of life, coevolution of organisms and environments, evolution of complex systems and their adaptations, evolution of ligand binding and catalytic function in proteins, evolution of a connected metabolism, evolution of patterns of gene regulation, evolution of development and its regulation, and so on. For a more popular presentation of the results of this book, see also Kauffman (1995).

[9] Cf., e.g., Paterson 1982. This line of work is discussed by Eldredge (1985, 1995); Eldredge and Grene (1992); Vrba (1980, 1984); and Vrba and Eldredge (1984).

[10] The theme of ahistorical universals is nicely developed by Kauffman. In light of his work, it is useful to distinguish between universals resting on fundamental physico–chemical laws and those resting on statistically near-universal features of the relevant classes of complex systems (e.g., those mentioned in note 8). The theme of constraints is usefully discussed in (Maynard Smith et al. (1985). In their terminology, constraints deriving from physico–chemical laws would be classed as "universal," those from the general features of complex systems as (relatively) "local."

its course. Nonetheless, the theory must also offer a means for weighing the causal relevance and relative importance of multiple processes, patterns, and singularities whose historical roles are not wholly determined by the background laws. For this reason, as well as others (e.g., the complications added by the hierarchical structuring of organisms and of evolutionary processes[11]), evolutionary theory is and will continue to be characterized by a proliferation of alternative models and smaller-scale theories applicable to particular cases. To this extent, the fundamental laws and principles of the theory cannot be expected to yield rigorous deductions of specific outcomes even when appropriate boundary conditions are supplied – although those outcomes *could* be derived from first principles plus boundary conditions *if only one causal process (or a small number of causal processes in a fixed relationship) were involved.*

Thus, basic evolutionary principles, even when supplemented with appropriate boundary or initial conditions, do not provide the wherewithal for a full derivation of major evolutionary patterns or the resolution of typical evolutionary disputes.[12] This is one reason for the seemingly inconclusive character of many debates over the dominant historical patterns in evolution (e.g., gradualism versus punctuation) and the mechanisms underlying those patterns (e.g., the causes of trends, the relative importance of selection and drift, the importance of ecological catastrophes, and the debates over the units of selection and the relevance of hierarchical structure). In all of these cases, we are dealing with questions of relative frequency. In all of these cases, examples can be found that support the existence of whichever pattern or the efficacy of whichever mechanism. And, in all of these cases, the patterns and mechanisms are compatible with the leading principles of the synthetic theory, *provided that those principles are stated abstractly enough.*[13]

[11] Hierarchical structure and its importance are discussed in passing in Burian (1988). A good start on the literature may be gotten from Eldredge (1985) and the following allied sources: Eldredge and Salthe 1984, Grene 1987, Salthe 1985, and Vrba and Eldredge 1984. These issues also connect in interesting ways with the huge literature on the units-of-selection problem.

[12] Cf. Brandon (1978, 1982b). Brandon argues that evolutionary theory gets its empirical content from the *specification* of these principles so that they pertain to particular cases (e.g., particular organisms and environments). The principles provide the schemata that, when specified, make empirical claims. It is for this reason that I count evolutionary theory as a schematic theory. Related positions, using various labels, have been taken by Caplan, Tuomi, Wasserman, and others; some references are supplied in Burian (1988).

[13] The *locus classicus* for such statements is Lewontin (1970). Lewontin's abstract version of the theory of natural selection can be boiled down to the following formula: heritable variation of fitness yields evolution by means of natural selection. This formula is compatible with group selection, non-Mendelian systems of inheritance, and even inheritance of acquired characters. To that extent, the formula seeks to represent the core of Darwinism, not the synthetic theory.

## PLURALISM

The distance from the abstract principles of variation, heritability, and differential fitness and the concepts on which they are founded to an account of the types and frequencies of the patterns of evolution and their causes is great indeed. Their very abstractness means that the principles are not sufficient, by themselves, to resolve the disputes like those just mentioned: The real work must go on closer to the data and with specific models. And, many of the models and scenarios in current use involve *specifications* of the abstract principles that depart in varying degrees from the spirit and content of the beliefs of the founders of the evolutionary synthesis. To this extent, even though no suitable radical alternative to the synthetic theory is in sight, *the fate of the synthesis as a coherent system of particular beliefs about evolutionary causes and patterns is still up in the air.*

This characterization may appear unduly pessimistic to some. I wish to counteract this appearance. On the one hand, the situation does not call for pessimism; new techniques now enable us to learn an immense amount that, until recently, was far beyond our means. On the other hand, the pessimism about the limited power of fundamental theories is, I believe, justified. Evolutionary biology is, unavoidably, a historical discipline, with a rich but still highly limited base of data available to it. Given this, it is unreasonable to expect to derive its principal results from a theory whose core consists solely of abstract or ahistorical (i.e., time symmetrical) laws. To put the point in a rather extreme way: I suspect that even such biologically basic matters as the specific content of the genetic code and the unique role of DNA in cellular organisms are contingent outcomes of historical processes. Given current knowledge, it seems unlikely that these fundamental properties of terrestrial organisms are necessary consequences of evolutionary laws applied to some class of carbon-rich planets that maintain, for a certain extended interval, a certain amount of surface water and a fairly temperate regime. Should this be correct, there can be little question but that the course of evolution, even on a fairly large scale, is fraught with the consequences of historical accidents and contingencies.

The point is not that ahistorical universals are irrelevant or unimportant – quite the contrary, for they, plus the relevant boundary conditions (e.g., those pertaining to the primordial earth), set the baseline for what would occur in the absence of selection or of any processes peculiar to living systems. Physical and chemical laws are causally prior to the origin of life. Other universals may pertain specifically to entities exhibiting certain of the structural complexities characteristic of living things. But, no matter: *All* relevant universals provide the setting within which the contingent history of living beings is played out. Just as the study of the geology of our planet cannot be properly pursued without reliance on both the fundamental laws of physics and chemistry as applied to planets and the specific accidental conditions pertaining to this planet, so the study of evolution generally and of the evolution of living forms on this planet cannot be properly pursued without reliance on the underlying laws of physics and chemistry plus any genuine laws pertaining to the features of complex systems of the types that happen to have evolved here. But, this leaves an important question open: *How much of the shape of the evolutionary history that we study has been determined by the accidents of circumstance (from molecular abundances at particular times and places to continental drift, volcanoes, and cometary impacts) that have impinged on the biota of the planet?* Such problems are found at all levels, from the molecular to the macroevolutionary.

To solve such problems, one needs to determine the "inertial baseline" from which biotic evolution and/or selection depart – for example, to determine what would happen to relevant sorts of complex genetic and biochemical systems – once they were up and running – in the absence of selection. A promising line of inquiry bearing on this topic has recently been opened up by Kauffman (cf. especially 1993). He shows how one can evaluate deep statistical features of very general classes of genetic systems so as to reveal important ensemble properties that would be manifested by genomes, proteins, cells, and organisms *in the absence of selection*. If this can be done, whether with Kauffman's protocols or with others currently being developed, it may be possible to make realistic estimates of the contribution of selection to the present genomic structures of organisms and of the extent to which similar structures should have been expected in the absence of selection.

More generally, various lines of work promise to yield improved estimates of the specific contributions of selection, drift, the structure of the environment, and rare catastrophes, as well as a better understanding of the patterns and structures that would arise no matter what as the automatic consequences of DNA and chromosomal mechanics and other structure-producing features of organisms. As a result, we can expect new data and new theoretical toeholds that can be put to use in evaluating the preponderance and importance of alternative modes and patterns of evolution and the various causes of those patterns. It is too early to tell whether this will enable us to resolve some of the long-standing issues that have plagued evolutionary theorists, but there are plenty of avenues to explore.

The apparent importance of the accidents of history, revealed by the fossil record and the distribution of properties among organisms, suggests that those accidents have played an enormous role in shaping the biota we study and the evolutionary patterns that we seek to understand. It is of great interest to learn to what extent this is true. Indeed, one of the most challenging intellectual problems posed by evolutionary biology is developing the proper tools to analyze the interplay between accident and law in shaping the familiar world around us.

## MORALS FOR THEORISTS AT THE CROSSROADS

It would be an act of hubris to try to predict the future of evolutionary biology in detail – especially so for an interested bystander like me. Furthermore, a serious estimate of where evolutionary theory is headed should examine a number of themes not addressed here. One of these concerns the role of hierarchies of various sorts in shaping evolutionary history; a second concerns the role of molecular work in transforming the practice of evolutionary biology. A third concerns the proper analysis of the logical structure of evolutionary theory and the ways in which it acquires its empirical content.[14] The views expressed in this section would be considerably enriched by developing these themes at length, but space prohibits opening up any additional topics. In any case, the two issues that dominate this chapter – the thoroughly interdisciplinary character of evolutionary biology and the historicity of the phenomena it studies – yield some ideas about the way things are likely to go. I conclude by offering a few of these ideas in the hope of provoking lively and thoughtful responses.

The historicity of evolutionary phenomena suggests certain limits on what can be expected from evolutionary theory. For this purpose, I include with evolutionary theory recent attempts to extend that theory or to connect it with other grand theories, like the attempts of Brooks and Wiley (1986) or Wicken (1987), for example, to connect the theory of evolution to non-equilibrium thermodynamics (NET) (cf. also Dyke 1987; Weber, Depew, and Smith 1988). NET is, of course, relevant to evolution, and it may tell us a fair amount about the character of the stable and self-perpetuating systems that are possible in a

[14] I find Brandon's argument, characterized in note 12, persuasive; the principle of natural selection, characterized in the text of this chapter, obtains its empirical content from the specifications of degrees of adaptedness for particular types of biological entities and types of environments.

wide range of circumstances. But, such considerations can capture at the very most certain baselines and thermodynamic constraints within which evolution occurs.

Lewontin suggested that the history of life can well be viewed as a history of the ways organisms have found to get around constraints (Maynard Smith et al. 1985, p. 282). Unless theories that characterize some constraints on organismic evolution (as NET does by studying "universal" constraints on the evolution of various kinds of dissipative structures) can also capture the role of particular historical circumstances in the breaking of constraints – and also, I would add, in determining lineage splitting, particular features of organisms or species, and the evolutionary effects of biological interactions and behaviors – those theories will not be able to capture the interaction between law and history that characterizes evolution.[15]

It is this interaction between law and history that requires any satisfactory general theory of evolution to have the peculiar character of a *schematic theory*.[16] By this, I mean that although such a theory provides a framework for describing and explaining evolutionary sequences and patterns, it is necessary to fill in that framework by means of particular empirical analyses of the historical features and circumstances of the organisms in question, including the peculiarities of their environment and the characteristics of traits and behaviors that will be advantageous in that environment. Thus, there is no character that is, of its own right, of high adaptive value; the adaptive values of all characters are relative to (partly accidental) historical circumstances.

It is here that the logical structure of the synthetic theory is so crucial. Brandon's analysis of that structure, which is consilient with the position I am now advocating, makes the point nicely. Brandon states the principle of natural selection as follows:

> (Probably) If *a* is better adapted than *b* in *E*, then *a* will have more offspring than *b* in *E* (Brandon 1982a, p. 432).

The interesting thing about this interpretation of the principle of natural selection is that, thus formulated, it provides no guidance regarding which traits or behaviors in which environments yield better adaptation than alternative traits. It is at just this point that empirical content is supplied to the principle by work undertaken in various independent disciplines, and it is at just this

[15] Kauffman's work, insofar as I understand it, has some chance of meeting these desiderata.
[16] See the references and discussion in note 12.

point that the relevance of the sorts of historical knowledge on which I have been focusing becomes inescapable.

If, indeed, historical and empirical work imported from a great variety of disciplines is required at this point, the character of the synthetic theory as a treaty is closely connected to its logically schematic structure. This structure, in turn, is tightly connected to the following four home truths: First, there are many ways that organisms can earn livings; second, an organism's success in earning a living depends not only on luck but also on its environment, including who its competitors are; third, no adequate answer as to how it is possible to earn a living can be derived primarily from general thermodynamic or evolutionary considerations, important as these are; and, fourth, there are ways of achieving reproductive success without being particularly good at making a living. If we take these home truths seriously, it should be clear that no single disciplinary matrix can provide satisfactory guidance regarding the issues raised within basic evolutionary theory. The lack of a single dominant disciplinary matrix in evolutionary biology is a consequence of the nature of evolutionary phenomena, and particularly of the role of historical accidents in affecting the evolutionary success, failure, and transformation of lineages. For this reason, I submit, we would be foolish to expect the unification of evolutionary theory within a single paradigm – and, what is more, we should count the failure to achieve such unification as a Good Thing.

## REFERENCES

Bowler, P. J. 1983. *The Eclipse of Darwinism.* Baltimore: Johns Hopkins University Press.

Brandon, R. N. 1978. "Adaptation and evolutionary theory." *Studies in the History and Philosophy of Science* 9: 181–206.

Brandon, R. N. 1982a. "A structural description of evolutionary theory." In *PSA 1980*, vol. 2, eds. P. Asquith and R. Giere. East Lansing, MI: Philosophy of Science Association, 427–39.

Brandon, R. N. 1982b. "The levels of selection." In *PSA 1982*, vol. 1, eds. P. Asquith and T. Nickles. East Lansing, MI: Philosophy of Science Association, 315–23.

Brandon, R. N. 1990. *Adaptation and Environment.* Princeton: Princeton University Press.

Brooks, D. R., and E. O. Wiley. 1986. *Evolution as Entropy: Toward a Unified Theory of Biology.* Chicago: University of Chicago Press.

Burian, R. M. 1988. "Challenges to the evolutionary synthesis." *Evolutionary Biology* 23: 247–69.

Darwin, C. R. 1859. *On the Origin of Species by Means of Natural Selection or the Preservation of Favoured Races in the Struggle for Life*. London: John Murray.

Depew, D. J., and B. H. Weber. 1988. "Consequences of nonequilibrium thermodynamics for the Darwinian tradition." In *Entropy, Information, and Evolution: New Perspectives on Physical and Biological Evolution*, eds. B. H. Weber, D. J. Depew, and J. D. Smith. Cambridge, MA: MIT Press, 317–54.

Dobzhansky, T. 1937. *Genetics and the Origin of Species*. New York: Columbia University Press.

Dyke, C. 1987. *The Evolutionary Dynamics of Complex Systems*. New York: Oxford University Press.

Eldredge, N. 1985. *Unfinished Synthesis: Biological Hierarchies and Modern Evolutionary Thought*. New York: Oxford University Press.

Eldredge, N. 1995. "Species, selection and Paterson's concept of the specific mate recognition system." In *Speciation and the Recognition Concept: Theory and Application*, eds. D. M. Lambert and H. G. Spencer. Baltimore: Johns Hopkins University Press, 464–77.

Eldredge, N., and M. Grene. 1992. *Interactions: The Biological Basis of Social Systems*. New York: Columbia University Press.

Eldredge, N., and S. N. Salthe. 1984. "Hierarchy and evolution." *Oxford Survey of Evolutionary Biology* 1: 184–208.

Glick, T. F., ed. 1974. *The Comparative Reception of Darwinism*. Austin: University of Texas Press.

Gould, S. J. 1983. "The hardening of the modern synthesis." In *Dimensions of Darwinism*, ed. M. Grene. Cambridge: Cambridge University Press, 71–93.

Gould, S. J. 1986. "Evolution and the triumph of homology, or why history matters." *American Scientist* 74: 60–9.

Grene, M. 1987. "Hierarchies in biology." *American Scientist* 75: 504–10.

Hull, D. L. 1985. "Darwinism as a historical entity: A historiographic proposal." In *The Darwinian Heritage*, ed. D. Kohn. Princeton: Princeton University Press, 773–812.

Jacob, F. 1977. "Evolution and tinkering." *Science* 196: 1161–6.

Kauffman, S. A. 1985. "Self-organization, selective adaptation, and its limits: A new pattern of inference in evolution and development." In *Evolution at a Crossroads: The New Biology and the New Philosophy of Science*, eds. D. J. Depew and B. H. Weber. Cambridge, MA: MIT Press, 169–207.

Kauffman, S. A. 1986. "A framework to think about evolving genetic regulatory systems." In *Integrating Scientific Disciplines*, ed. W. Bechtel. Dordrecht, Holland: Nijhoff, 165–84.

Kauffman, S. A. 1993. *Origins of Order: Self-Organization and Selection in Evolution*. New York: Oxford University Press.

Kauffman, S. A. 1995. *At Home in the Universe: The Search for the Laws of Self-Organization and Complexity*. New York: Oxford University Press.

Kellogg, V. L. 1907. *Darwinism To-Day*. New York: Holt.

Kohn, D., ed. 1985. *The Darwinian Heritage*. Princeton: Princeton University Press.

Kuhn, T. S. 1962. *The Structure of Scientific Revolutions*. Chicago: University of Chicago Press.

Kuhn, T. S. 1970a. "Reflections on my critics." In *Criticism and the Growth of Knowledge*, eds. I. Lakatos and A. Musgrave. Cambridge: Cambridge University Press, 231–78.

Kuhn, T. S. 1970b. *The Structure of Scientific Revolutions*, 2nd ed. Chicago: University of Chicago Press.

Kuhn, T. S. 1974. "Second thoughts on paradigms." In *The Structure of Scientific Theories*, ed. F. Suppe. Urbana: University of Illinois Press, 459–82.

Lewontin, R. C. 1970. "The units of selection." *Annual Review of Ecology and Systematics* 1: 1–18.

Lewontin, R. C. 1983. "The organism as subject and object of evolution." *Scientia* 118: 63–82.

Masterman, M. 1970. "The nature of a paradigm." In *Criticism and the Growth of Knowledge*, eds. I. Lakatos and A. Musgrave. Cambridge: Cambridge University Press, 59–89.

Maynard Smith, J., R. M. Burian, S. A. Kauffman, P. Alberch, J. H. Campbell, B. Goodwin, R. Lande, D. M. Raup, and L. Wolpert. 1985. "Developmental constraints and evolution: A perspective from the Mountain Lake conference on development and evolution." *Quarterly Review of Biology* 60: 265–87.

Mayr, E. 1942. *Systematics and the Origin of Species*. New York: Columbia University Press.

Mayr, E. 1980. "The role of systematics in the evolutionary synthesis." In *The Evolutionary Synthesis: Perspectives on the Unification of Biology*, eds. E. Mayr and W. B. Provine. Cambridge, MA: Harvard University Press, 123–36.

Mayr, E. 1982. *The Growth of Biological Thought: Diversity, Evolution, and Inheritance*. Cambridge, MA: Harvard University Press.

Mayr, E. 1985. "Darwin's five theories of evolution." In *The Darwinian Heritage*, ed. D. Kohn. Princeton, NJ: Princeton University Press, 755–72.

Paterson, H. E. H. 1982. "Perspectives on speciation by reinforcement." *South African Journal of Science* 78: 53–7.

Salthe, S. N. 1985. *Evolving Hierarchical Systems*. New York: Columbia University Press.

Seward, A. C., ed. 1909. *Darwin and Modern Science*. Cambridge: Cambridge University Press.

Simpson, G. G. 1944. *Tempo and Mode in Evolution*. New York: Columbia University Press.

Stebbins, G. L. 1950. *Variation and Evolution in Plants*. New York: Columbia University Press.

Todes, D. P. 1987. "Darwin's Malthusian metaphor and Russian evolutionary thought, 1859–1917." *Isis* 78: 537–51.

Todes, D. P. 1988. *Darwin without Malthus: The Struggle for Existence in Russian Evolutionary Thought*. New York and Oxford: Oxford University Press.

Vrba, E. S. 1980. "Evolution, species and fossils: How does life evolve?" *South African Journal of Science* 76: 61–84.

Vrba, E. S. 1984. "Evolutionary Pattern and Process in the Sister-Group Alcelaphini-Aepycerotini (Mammalia: Bovidae)." *Living Fossils*: 62–79.

Vrba, E. S., and N. Eldredge. 1984. "Individuals, hierarchies, and processes: Towards more complete evolutionary theory." *Paleobiology* 10: 146–71.

Wallace, B. 1989. "Populations and their place in evolutionary biology." In *Evolutionary Biology at a Crossroads*, ed. M. K. Hecht. New York: Queen's College Press, 21–44.
Weber, B. H., D. J. Depew, and J. D. Smith, eds. 1988. *Entropy, Information, and Evolution: New Perspectives on Physical and Biological Evolution*. Cambridge, MA: MIT Press.
Wicken, J. S. 1987. *Evolution, Thermodynamics, and Information: Extending the Darwinian Program*. New York: Oxford University Press.

# 6

# "Nothing in Biology Makes Sense Except in the Light of Evolution" (Theodosius Dobzhansky)[1]

Theodosius Dobzhansky (1900–75) was one of the principal architects of the so-called synthetic theory of evolution.[2] Born and initially trained in Russia, where he became an entomologist and zoologist with wide-ranging interests, he brought a rich background in systematics and study of natural populations with him when he came to the United States to learn genetics in the laboratory of T. H. Morgan in 1927 (Adams 1994). In the end, he remained in the United States, in part because of political developments in Russia, where he was ultimately declared a nonperson. Dobzhansky utilized his double background of work with natural populations and in Mendelian genetics in writing what came to be the singlemost influential book in the formative period of the synthetic theory of evolution (Dobzhansky 1937). Dobzhansky is a particularly interesting figure to study because of the cultural dualities (or, rather, multiplicities) that he embodies, because his work forcibly reconciled biological disciplines (i.e., systematics and natural history on the one hand, genetics on the other) that had not only drifted apart but had also built strong evidentiary bases for conflicting claims, and because – like many of the founders of the evolutionary synthesis – he sought to relate biological findings to issues in the wider culture.

[1] The first version of this chapter was presented at a colloquium in honor of Marjorie Grene, "Conceptions de la science: hier, aujourd'hui et demain," that Jean Gayon and I organized at the University of Burgundy in May 1995. The present version has been greatly improved by discussion at the colloquium and the criticisms and suggestions of Jean Gayon and Marjorie Grene. A French version of this paper is scheduled to be published in 2005 in *Conceptions de la Science – hier, aujourd'hui et demain: hommage à Marjorie Grene*. Jean Gayon and Richard Burian, eds., with the collaboration of Marie-Claude Lorne. Brussels: Ousia.

[2] About the synthetic theory and some of the current debates about its status, see, e.g., Burian (1988); Gayon (1990); Grene (1983, 1990); and Mayr and Provine (1980). For further biographical details about Dobzhansky, see Adams (1994), Ayala (1990), Gould (1982), and Provine (1981).

This chapter takes its title from a talk that Dobzhansky delivered to the American Association of Biology Teachers (Dobzhansky 1973a). The title of that talk, "Nothing in Biology Makes Sense Except in the Light of Evolution," is often cited both in controversies over the status of evolutionary studies within biology and in various cultural debates over the (supposed) conflict between evolution and religion. In this chapter, I examine Dobzhansky's claim and some of his arguments in its favor with an eye to both of these contexts. I focus first on the setting – both within biology and within American culture – of Dobzhansky's arguments for the importance of evolutionary biology. Second, I turn to the relations among biological disciplines or fields, especially between evolutionary biology and molecular genetics – a field that "took off" after the formation of the evolutionary synthesis. Finally, I deal with some complications arising from the difference between the intellectual and the institutional positions of the field of evolutionary biology. I can, of course, do only partial justice to Dobzhansky's advocacy of the centrality of evolutionary studies within the biological sciences.

## EVOLUTIONARY THEORY IN BIOLOGY AND EVOLUTIONARY BIOLOGY IN THE UNITED STATES

A remarkable fact about evolutionary biology, given its importance in twentieth-century biology, is that there were virtually no departments of evolutionary biology for most of the century (as there have been of genetics, for example).[3] It is important to start from the fact that Mendelian genetics and Darwinian evolutionary biology were in serious conflict with each other in the first third or so of the last century. There are many aspects to this conflict, both intellectual and institutional.[4]

[3] Interestingly, the major exception is Russia where, early in the century, there were intense and productive scientific debates over various Darwinian and non-Darwinian theories of evolution. At the end of the 1930s, most major Soviet universities had departments of Darwinism. But, in consequence of Russia's intellectual isolation in the 1930s, the controversies over Lysenkoism, and Lysenko's ascent to power, they deployed a distinctive Soviet version of Darwinism that did not match orthodox evolutionary theories and practices in the West at that time. On this topic, see Adams (1991). See also Provine (1980) for some useful material on the institutional bases for genetics. Provine does not explicitly take up the formation of departments of genetics within colleges and universities.

[4] There is a huge literature on the debates involved. Important references include Bowler 1983, 1988, 1989; Depew and Weber 1995, Part II; Gayon 1992, 1998; and Provine 1971. On this last topic, for a contrasting point of view, see Nordmann (1994). In a recent paper, Gilbert ("Back to the Future: Resynthesizing Evolutionary and Developmental Biology," presented at the Fifth

I offer here a first approximation account of some of the differences involved. Some of them follow from the practical necessity for early Mendelians to work with mutations with large effects. Indeed, because at first all known mutations were large in effect, the Mendelians held that such mutations were the norm and provided the basis on which new species were formed. In contrast, most natural historians and Darwinians insisted on the absence of large mutations in the field and on the gradual character of changes from one species to another. Thus, early Mendelians and turn-of-the-century Darwinians agreed that the effects of Mendelian mutations were too large to allow natural selection on Mendelian mutations to control the features of organisms with any precision, including features of organisms in a new species. Natural historians did not find significant numbers of Mendelian mutants in natural populations; accordingly, the Darwinians derided Mendelian mutations as experimental freaks that would be eliminated immediately from natural populations and as irrelevant to evolutionary history. Geneticists, in contrast, argued that Darwinians had no adequate causal basis from which to explain the changes that took place in speciation and that were marked by the striking visible differences between "good species."

Turning to the institutionalization of biology, the term "biology" came to be employed rather self-consciously in some quarters around the turn of the century for *experimental* biology. The insistence on experiment was used to help legitimize a series of new sciences – genetics among them – as distinct from natural history, botany, zoology, systematics, paleontology, and speculative Darwinian evolutionism. Part of the point of excluding these old-fashioned sciences was that they did not have adequate means of testing explanatory hypotheses or, worse yet, they did not even offer genuine explanatory hypotheses. One aspect of the movement toward experimental biology was the physics envy of biologists and their long struggle to demonstrate that their science is as legitimate as physics, a struggle in which Darwin himself had been engaged. Many participants in the experimental movement deliberately

Mellon Workshop on History and Social Sciences of Contemporary Life Sciences, MIT, 1995) points out an important institutional aspect: During the 1950s, population genetics received considerable funding from the U.S. Atomic Energy Commission in connection with concerns about the effect of radiation on populations. As he puts it: "Whereas most evolutionary studies have difficulty getting funds and students, concerns about the genetic effects of radiation enabled Dobzhansky and others a constant supply of money and graduate students" (MS p. 5). On this topic, see also Beatty (1991) and his "Opportunities for Genetics in the Atomic Age," presented at the Fourth Mellon Workshop on History and Social Studies of Contemporary Life Sciences, MIT, May 1994. The lines of work thus funded thrived, but they did not have significant impact on the organization of academic departments or on the larger theoretical issues in evolutionary biology.

excluded Darwinian theory from (their kind of) biology precisely because it was "too speculative" and had no means of adequately controlling or testing its speculations (especially after the impact of Haeckel and Weismann).[5] The success of this "experimentalist" movement is marked internationally by the dominance of the new experimental disciplines in the departments and research units begun after, say, 1910.

Although this is only an approximate account of one part of the deep background to the matters of primary concern here, it usefully marks Dobzhansky's accomplishment. During the 1920s and into the 1930s, the founders of mathematical population genetics – especially Fisher (1930), Haldane (1932), and Wright (1931, 1932) in the Anglophone world[6] – using models employing large numbers of mutations with small effects, provided mathematical demonstrations that Mendelism is compatible with Darwinian gradualism. Indeed, given enough mutations of small effect, large-enough populations, stable-enough conditions, and enough time to approach equilibrium, the mathematics demonstrated that the selection would be the dominant factor in shaping the changing characteristics of organisms within a population. At the same time, laboratory work in Mendelian genetics had demonstrated that there are many more mutations with small effects than mutations of large effect and that the mutations of small effect can modify what is done by those with large effects. Thus, the groundwork was laid for reconciling genetics with studies of natural populations. But, the relevant communities of investigators were not sympathetic with one another and did not read each other's work. And, there were certainly plenty of serious questions to be asked about whether the assumptions of the models corresponded to the biological realities to be found in natural populations and about the relevance of change within populations to speciation. It was Dobzhansky's virtue to bridge the communities of naturalists and geneticists, find a methodology for doing experimental work with natural populations, and write a book that formulated the issues so well that, after much dispute, both communities were, by and large, persuaded.

In part on the basis of his extensive work with natural populations, Dobzhansky had always held that there is an immense reservoir of variation within populations in nature. In *Genetics and the Origin of Species* (1937), he synthesized his findings and those of many others to provide an integrated account, consistent with genetics, of the origin of species as an extrapolation

[5] Mark Adams forced me to recognize the importance of the experimentalist movement. I am grateful to him for many helpful discussions of this topic. A major debate on this issue was triggered by Allen (1979) and a symposium on the topic edited by Jane Maienschein, Ronald Rainger, and Keith Benson, *Journal of the History of Biology 14* (1981), pp. 83–176.

[6] Dobzhansky, of course, also knew the work of Chetverikov and Timofeeff-Ressovsky.

of the microevolutionary changes that the naturalists had described in detail. He also argued that those results are entirely consistent with the models of population genetics (particularly Wright 1931, 1932). Furthermore, his account did not employ the difficult mathematics of the population geneticists. In consequence, it made their results accessible to both the larger genetics community and the natural historians, who were largely unequipped to handle the mathematics (as was the case with Dobzhansky himself). In the early papers of a series of forty-three experimental papers on the genetics of natural populations, the first eighteen of which were published from 1938 to 1948 (Lewontin et al. 1981), he managed to show that natural populations have a large reservoir of (mostly recessive) Mendelian mutations, that there are significant genetic differences between local populations, and that the predictions of certain of Wright's mathematical models are compatible with the changes found in the populations of *Drosophila pseudoobscura* that he studied.

It is worth noting a controversial problem on which Dobzhansky took a strong stand. The issue stems from a paradox: The more powerful natural selection is, the more likely it is to consume the variation that it requires to change the characteristics of the organisms in a population. The reason is simple: Where there are selective differences among conspecific organisms within a population, there is some likelihood of losing traits that are less advantageous in favor of a single optimal solution to the physiological and ecological problems faced by the members of the population (or the closest to optimal among the traits that are available). The stronger the coefficients of selection, the more rapidly alternative traits are lost. So how can there be high degrees of variation within a population if the traits of the organisms in that population are shaped by selection? To the extent that the proponents of the synthetic theory – Dobzhansky among them – hold that the dominant factor shaping organisms in evolution is natural selection, they should expect variation to be minimized. How can Dobzhansky's insistence on the power of selection in leading to the origin of species be reconciled with the high degrees of variation he found, required, and advocated as the typical condition of natural populations?

Many geneticists (e.g., Herman Muller, 1950, who occupied the "classical" position) applied this argument at the genetic level. They argued that selection tends to *reduce* genetic variation and that a population near equilibrium will be largely homozygous, with only an occasional mutant allele. Dobzhansky, in contrast, advocated what came to be known as the "balance" position: At equilibrium, most loci would be heterozygous, that there were many more than two alleles available within the population at most loci, and that the action of the genes will be balanced in such a way that the external phenotype will be rather homogeneous, masking the enormous amount of genetic variation within the population.[7] To support this view, Dobzhansky argued that the availability of different alleles enables organisms to respond successfully to the wide range of environmental conditions that they encounter in their lifetime.[8] Putting it crudely, selection favors phenotypes capable of responding to a wide range of humidities, temperatures, constituents in food, toxins, and so on, and to adjusting developmental schedules to meet environmental conditions. The best way to do this is to have (to use Waddington's term) a *canalized* phenotype[9] plus high amounts of genetic variation, so that most individuals have two distinct alleles available at most loci, unless there is specific harm done by one of those alleles (see Waddington 1942, 1960).

This argument depends crucially on denial of the assumption that genes are "the target of selection" (a term coined by Mayr).[10] This strategy is natural for those who study natural populations, for they understand selection in terms of the differential consequences of possessing different *phenotypes*; however, it is less natural for geneticists, who calculate selection coefficients wholly in terms of *changes of gene frequencies*.[11] Curiously, Dobzhansky

[7] There is continuing controversy about the extent of justification for and truth in these two polar positions. During Dobzhansky's lifetime, the epistemological difficulties involved in resolving the dispute were insuperable; they remain severe even now. The *locus classicus* for setting out the biological side of the epistemological issues is Lewontin (1974). Lewontin's analysis of the subsequent ramifications of the controversy is controversial (see, e.g., Dietrich 1994).

[8] I have ignored some important shifts in Dobzhansky's position over time. Two issues on which his views changed concern the extent of coadaptation among the alleles represented in a given local population (see Lewontin 1993) and the importance of the effects of small population size on the variation within populations. (This factor made an essential contribution in Wright's models of evolutionary change.) As Dobzhansky became more selectionist and the synthesis "hardened" (cf. Gould 1983a), random variation in small populations came to play less of a role in Dobzhansky's and many other synthetic theorists' accounts of evolution.

[9] This means, roughly, that the development of the organism is so programmed that it lands in more or less the same adult (and intermediate) external phenotypes despite wide variation in environmental conditions. For an early text setting forth this important concept, see Waddington (1942).

[10] See, e.g., Mayr 1959, 1962, and 1963. The argument also depends on assumptions about mechanisms of gene expression.

[11] Mayr has repeatedly criticized Dobzhansky and others for *defining* evolution in terms of change of gene frequencies. There are many additional reasons for resisting such a definition – for example, the way in which it underemphasizes the centrality of speciation (which cannot be defined in terms of gene frequencies) in evolutionary change. The issue of the extent to which *natural selection* can be understood primarily by means of calculations at the genic level continues to be actively debated – for example, in controveries over the *genic selectionism* of R. Dawkins, G. C. Williams, and others. See, for example, Dawkins (1976, 1982) and Williams (1966). There are a number of important connections to some issues about biological hierarchies – see, for

himself adopted the formula that evolution equals change of gene frequency, a position for which Mayr severely criticized him.

Turning briefly to larger contexts, two topics serve as emblems of the general situation in the United States: (1) the institutionalization of research support within biology, and (2) the tendency in American society to see evolutionary biology as standing in conflict with organized religion and with creationist doctrines. These two issues are intertwined to a surprising degree. One way of showing the connection is to reflect on the teaching of biology in U.S. high schools. Gould did so in an essay written in December 1981 while attending a trial in which a law from the state of Arkansas was being challenged (1983b, Chapter 21). The law, ultimately declared unconstitutional, schools to give equal treatment to required "Creation Science" and "the Evolution Theory" (supposedly equally justified and scientific). Gould's essay descries the lack of evolutionary content in the dominant high school biology textbook, from which both he and I were taught in the 1950s. The book, *Modern Biology* by Moon, Mann, and Otto, was the lineal descendant of *Biology for Beginners* by Truman J. Moon, first published in 1921. Gould points out that the frontispiece of the original text was an image of Charles Darwin and that the book was organized around the idea of evolution. He shows that after 1925, the date of the infamous Scopes trial, the evolutionary content was removed from the text. Not only is Darwin's image gone but also the very term "evolution" is replaced by the cowardly paraphrase, "the hypothesis of racial development"! The 1956 edition, for example, contains a brief chapter (the 58th of 60) touching on evolution. The chapter evades the issue of creation versus evolution by such circumlocutions as the following:

> During these ages [of geological change], species of plants and animals have appeared, have flourished for a time, and then have perished as new species took their places. . . . When one race lost in the struggle for survival, another race appeared to take its place.

This is how high school biology was taught to my generation. This sad fact reflects the political power of biblical fundamentalism in the United States and the country's genius for intellectually unsatisfactory compromise. Against this background, is it any wonder that there was no impetus, even among professional biologists, to place evolution at the center of biology curricula or to organize biological disciplines so as to ensure that evolutionary studies played a key role in the organization of biological work in general?

example, Grene 1987, especially the relations of ecological and genealogical hierarchies and various debates over the level(s) of selection.

Dobzhansky was keenly aware of this situation and faced it head-on in the address whose title is the name of this chapter.

Although it is not simple to demonstrate, it is extremely plausible that this cultural background had an important influence on the structure of professional biology. Gilbert, for one, connects the strength of creationism to the structure of the National Science Foundation (NSF): "In the United States, evolution is still so suspect that no National Science Foundation study section is designated evolutionary biology."[12] Difficult as it is to make a direct connection to the cultural milieu in such a matter, I believe that there is widespread consensus among evolutionary biologists that this has been a central problem in the institutionalization of their (would-be?) discipline. The contrast with genetics, which receives major public and private funding in virtue of its economic importance, is striking.

## DOBZHANSKY'S ARTICLE[13]

Dobzhansky begins his article by discussing the rejection, on scriptural grounds, of the Copernican theory of the heavens as a "mere theory," not a "fact" by a Saudi fundamentalist, Sheik Abd el Aziz bin Baz. Dobzhansky lays out a standard argument against this view in a Darwinian style, arguing that the hypothesis that the earth orbits the sun "makes sense of a multitude of facts that are otherwise meaningless or extravagant." Dobzhansky is ready to grant, for the sake of the argument, that no direct observations demonstrate the "fact" that the earth orbits the sun but insists that we must dismiss the geocentric alternative as making nonsense of many facts we *have* observed. Similarly, there are no direct observations of the age of the earth or of the evolution of new species, but an enormous multitude of facts shows both the diversity and the unity of life that would be meaningless or extravagant except in the light of evolution.

Rhetorically, Dobzhansky makes excellent use of the parallels between the pro-Copernican and pro-Darwinian arguments and the fact that the anti-Copernican fundamentalist is a Muslim from Saudi Arabia. By adding the cultural distance between a Saudi fundamentalist and mainstream Americans to the strongly held belief in a heliocentric solar system in the United

[12] This quotation was taken from an unpublished manuscript that I can no longer locate. For relevant background and data, see Smocovitis (1996, Chapter 3, esp. pp. 65–72).

[13] Throughout this section, page numbers refer to Dobzhansky (1973a).

States, Dobzhansky makes it difficult for a biblical literalist to contest the pro-Copernican argument. Dobzhansky argues that the literalist is put in the position of blaspheming if he treats the Koran or the Bible as a primer of natural science, for when scientific developments outrun the literal reading of scripture, the literalist is forced to deny well-established facts or to make God into a deceiver by planting misleading evidence.[14] Beyond that, he argues, scripture is meant to deal with far more important matters, "the meaning of man and his relation to God," matters that require the use of poetic symbols comprehensible both to the people of a given era and to people from other times (p. 125). In the main body of the paper, Dobzhansky spells out the parallel argument illustrating at some length the multitude of facts in biology (but also geology, radiometric dating, and much more) that evolution is required to explain.

Of particular interest here is Dobzhansky's treatment of molecular biology. The diversity of life is quickly evident from the sheer number of species and bodily diversity among plants, animals, and microorganisms; the extraordinary diversity of the ecological niches they occupy; and the narrowness of the adaptations that many organisms exhibit. It is somewhat harder to establish the unity of life. Dobzhansky treats molecular findings as the most decisive but by no means the only evidence for the unity of life. (He also cites, for example, traditional evidence of homologies and of shared developmental pathways from comparative anatomy and embryology and illustrates the ways in which diversity can come from unity by the remarkable radiation of *Drosophilid* flies on the Hawaiian Islands.[15]) Molecular evidence for the unity of life includes the universality of DNA and RNA as the genetic materials; the universality of the genetic code and the mechanisms by means of which the code is translated into sequences of amino acids; the high uniformity of cellular metabolism (cf. the roles played by compounds such as adenosine triphosphate, hemes, riboflavin, and pyridoxin in nearly all organisms); and

[14] For example, in dealing with the massively coherent radiometric evidence of the age of the earth, dismissed as "mere theory" by "Sheik bin Baz and his like," one must ask, "What is the alternative? One can suppose that the Creator saw fit to play deceitful tricks on geologists and biologists. He carefully arranged to have various rocks provided with isotope ratios just right to mislead us into thinking that certain rocks are 2 billion years old, others 2 million, while in fact they are only some 6,000 years old" (p. 126).

[15] We now know that more than 800 of the world's more than 2,000 species of *Drosophila*, including some of the most aberrant ones, are found on a geologically isolated landmass 3 percent the size of France. Because of their volcanic origin, the islands have well-defined ages; the oldest is less than 6 million years old, the youngest less than 1 million years. The immense diversification and speciation of a founding stock must have taken place in that time.

the consistency of the rate of substitution of amino acids into such highly conserved proteins as the cytochromes. Such

> biochemical or biological universals . . . suggest that life arose from inanimate matter only once and that all organisms, no matter how diverse in other respects, conserve the basic features of the primordial life. (It is also possible that there were several . . . origins of life; if so, the progeny of only one of them has survived and inherited the earth.) But what if there was no evolution and every one of the millions of species was created by separate fiat? However offensive the notion may be to religious feeling and to reason, the antievolutionists must again accuse the Creator of cheating. They must insist that He deliberately arranged things exactly as if his method of creation was evolution, intentionally to mislead sincere seekers of truth (p. 127).

The claimed centrality of evolution within biology was meant both for biologists and for a larger public. This is demonstrated by the whole tenor of Dobzhansky's career. To cite but one piece of evidence regarding his interest in the larger public, one need only look at the books he published aimed at a wide audience.[16] Within biology, I note his constant efforts to place all biological questions within an evolutionary framework and to ensure an institutional footing (and funding) for evolutionary studies. It is a minor but indicative symptom that the phrase "nothing in biology makes sense except in the light of evolution" serves as the epigraph of a coauthored textbook on evolution, published posthumously (Dobzhansky et al. 1977).

## MORE ON THE RELATIONS AMONG BIOLOGICAL DISCIPLINES

I now leave Dobzhansky to examine some complexities concerning the relationship between evolutionary studies and molecular biology. There is considerable irony in the interactions between the two, especially with regard to the stance that Dobzhansky and other leading "nonmolecular" evolutionists took in favor of the centrality of evolution within biology. I discuss the relevant developments briefly and abstractly, first in terms of the relationships among biological disciplines, institutions, and funding, then in terms of some issues about the molecularization of biology.

An interesting way of thinking about the history of biology in the twentieth century is in terms of the founding of moderately autonomous disciplines

[16] The list includes Dobzhansky 1956, 1963, 1964, 1967, 1973b; Dobzhansky and Dunn 1946; and Dobzhansky, Boesiger, and Wallace 1983. Dobzhansky's activity along these lines includes a considerable amount of periodical literature and editorial work as well.

and their subsequent interactions. Thus, as indicated in the first section of this chapter, early Mendelian genetics was fairly sharply separated from (Darwinian) evolutionary biology. An enormous gulf also arose between genetics and embryology. The two disciplines employed different techniques, came to adopt conflicting experimentally grounded presuppositions (e.g., concerning the nuclear versus cytoplasmic origin or control of key traits of multicellular organisms), and employed different organisms (with strikingly different properties) for the exemplary research on which disciplinary consensus was built.[17] Specialization became ever more important in the course of the twentieth century, and it brought with it the price of significant barriers among biological disciplines, both within the experimental tradition and between experimentalists and naturalists. There were, of course, always those who sought to integrate the findings (and occasionally the techniques) of different specialties, but this was no simple matter. It is, of course, the job that Dobzhansky believed could be accomplished by evolutionary biology. But, insofar as the institutionalization of biology is concerned, as already indicated, no firm institutional setting was developed to serve as a base for evolutionary biology as an integrative discipline, at least into the 1980s.

The roots of molecular biology (which are currently much disputed)[18] are to be found in transmission genetics, crystallography, information theory, biochemistry, the application of certain techniques from physics, and a number of allied sources, none of which are centrally interested in evolutionary questions. I have argued in a number of articles (as have many others) that molecular biology is best understood as a loosely interrelated collection of tools, wedded to the ideal that one can elucidate and clarify biological function by analyzing molecular structures.[19] To the extent that this characterization is fair, at least as a first approximation, the problems of evolution do not belong to the initial core problems of molecular biology; although, of course, someone interested in evolutionary questions can apply the new molecular tools to them – something that has been done with a vengeance in recent years with the development of a major (sub?)discipline of molecular

[17] On this topic, see Chapter 10 or Sapp (1987). For one embryologist's view of the threat posed by genetics to his discipline, see Harrison (1937). For my views on some more recent developments, see Chapters 11 and 12.

[18] The broadest standard history is Judson 1996 (1979). For an unusual position on the proper delimitation of molecular biology, see Zallen (1993). Morange provides a useful synthesis of conflicting views on this topic, including a criticism of Zallen's point of view in Morange (1994, 1998).

[19] See, for example, the symposium on "The Tools of the Discipline: Biochemists and Molecular Biologists" (de Chadarevian and Gaudillière 1995). I state some of the points at stake here more forcefully in Burian (1996).

evolution and the flourishing of evolutionary studies within a number of the specialties of molecular biology. Nonetheless, as I argue in the concluding section of this chapter, the pathway by means of which the development of studies of molecular evolution came into being was neither straightforward nor obvious.

## MOLECULAR BIOLOGY AND EVOLUTION

There are two aspects to the difficult birth of molecular evolutionary studies: one is the weak institutionalization of evolutionary studies in general; the other concerns some aspects of the intellectual history of early work in molecular biology. Institutionally, as I argued previously, there were few natural bases for evolutionary biology, especially in connection with molecular biology, rooted as it was in the experimentalist traditions that tended to exclude evolution as a serious topic of study. As I did not argue explicitly, early molecular biology drew heavily on biochemical and molecular studies and other lines of work initiated by the "invasion of the physicists."[20] In spirit, it was experimental and mechanistic. It required experimental demonstration of the specific action of particular genes or physical–chemical influences, such as particular radiations, chemical mutagens, and metabolites. One characteristic of the new generation of molecular biologists was the drive toward quick and decisive experimental resolution of experimentally tractable questions. This meant that they favored the use of microorganisms in virtue of their experimental tractability and the rapidity of the experiments they permitted. Indeed, studies based on microorganisms displaced slower evolutionary projects both in funding and in the postwar institutionalization of biology. Studies of long-term evolutionary scenarios were looked on unfavorably, whereas short-term mechanistic questions, for which rapid techniques were available (cf. ultracentrifugation, electrophoresis, use of radioactive tracers, plating of phages), came to dominate molecular genetics.

Work on the mechanisms of molecular action displaced work with traditional genetic and evolutionary organisms (and was favored with better funding) in part because of the obvious potential for applications to human health and improvement of agricultural production. The diversion of funding to these purposes tended to displace more general evolutionary studies. The reshaping of departments of biology in the postwar years followed the same trends. Accordingly, in many leading institutions, evolutionists felt that

[20] Two classical sources on this topic are Fleming 1968 and Judson 1996 (1979).

they were fighting rear-guard actions to retain their positions and prestige in the face of the expansion of molecular biology and the indifference of most molecular biologists to evolutionary issues.

One particular consequence of the concentration on microorganisms in the early days of molecular biology is particularly ironic. Because microorganisms (especially viruses and procaryotes – organisms with no true nuclei and relatively limited amounts of DNA) experience enormous pressure for economy in the use of genetic material, they are under much greater pressure to find optimal solutions to environmental problems than are eucaryotes (nucleated organisms, mostly multicellular, with excess DNA buffered by a variety of protective devices).[21] Again, because of lack of buffering of the genome, there is a tendency for nonfunctional genetic variation in microorganisms to be more readily eliminated than in eucaryotes.[22] As a result, many geneticists who worked with phage and bacteria were prepared to infer that the organisms they studied were "perfectly adapted"[23] or nearly so. What this means is that microorganisms were perceived as having eliminated genetic variants that interfered with optimal performance in physiological conditions. In this respect, they do not carry obvious stigmata of their history; rather, adaptation dominates over historical pathways. Thus, the early findings in microbial genetics tended to find relatively less variation that Dobzhansky expected and,

[21] Including diploidy; many of the microorganisms studied were haploid and so each and every allele was required to function properly for survival – a claim that is not true for diploid organisms.

[22] Among the many reasons for this, eucaryotes are typically diploid. Many genes that would be lethal if they were present in a single copy or when homozygous are harmless or even beneficial when heterozygous because they are protected by being paired with a functionally normal allele. Because of interactions among genes and gene products, the heterozygotes containing such "lethal" genes can even be beneficial, as in the example of sickle cell anemia. Again, eucaryotes often have highly evolved systems of gene families. Multiple copies of physiologically important genes allow some copies to evolve silently because other copies of the gene perform the required function. Similarly, other features of eucaryotic genomes can increase the protection of a particular gene from the immediate action of selection. For example, the separate pieces of "interrupted genes" in eucaryotes (which have separated units, often corresponding to functional parts of the corresponding protein) can be duplicated independently of one another and combined with pieces from other genes. This allows *parts* of genes to be stabilized and recombined and to evolve independently of one another. Yet, again, eucaryotes often do a great deal of "post-transcriptional modification" of messenger RNA. That is, they have multiple means of processing genetic information after it is transcribed onto RNA and before it is "translated" (i.e., decoded to yield a protein). These devices allow evolution of the *modulation* of a genetic signal without a direct change in a particular gene. There are literally hundreds of complex phenomena that could be listed here.

[23] For development of the importance of the doctrine of perfect adaptation in Darwin's work and the struggle to replace it with the weaker notion of relative (sufficient, but imperfect) adaptation, see Ospovat (1981). For one digest of the importance of this shift for the evolutionary synthesis, see Chapter 4.

thus, to support the notion of "perfect adaptation" at equilibrium and with it the classical rather than Dobzhansky's balance theory. These conclusions were, of course, resisted by those evolutionists who worked with natural populations, especially populations of diploid eucaryotes. Such naturalists had long since developed evidence for the presence of large amounts of covert variation in populations of eucaryotic organisms.

In the mid-1960s, the biochemical technique of gel electrophoresis was adapted for use in population studies of multicellular eucaryotes.[24] These techniques could and did detect enormous amounts of variation in the amino-acid composition of particular proteins in natural populations. These new molecular techniques were deployed to address the question of the amount of variation in diploid populations and thereby, indirectly, the potential for evolutionary change in seemingly monomorphic populations. The results led to immense and continuing controversy. However, to a first approximation, it is safe to say that they supported Dobzhansky's balance against the classical position at least in this respect, that they showed there to be immensely more variation present in natural populations than could readily be accounted for on the classical hypothesis. Precisely how these findings affect the debates about the power of natural selection and what bearing they have on the paradox caused by the tendency of natural selection to consume the variation that is relevant to the formation of an optimum remain a matter of controversy to this day.[25] This is a story that we cannot follow here, but one that allows us to bring to closure this brief study of the status and influence of Dobzhansky's stance about the importance of evolution within biology.

## CONCLUSION

During the last thirty years, molecular biology has produced a series of extremely startling findings. As was once said by astronomers of the system of the heavens, if God had consulted the biologists, they would have been able to devise for Him simpler ways of building the systems that maintain life. In particular, organisms are built of a variety of pieces and systems that were not designed to fit together. As is now well known, this is true down to the molecular scale. Genes of eucaryotes are built in pieces that have to be spliced together. The intermediate products encoded by those genes are

[24] The breakthrough papers, which set off an entire industry, were Hubby and Lewontin (1966) and Lewontin and Hubby (1966).

[25] The *locus classicus* for discussion of this debate is Lewontin (1974). See also Dietrich (1994).

sometimes altered as much as six or seven times at six or seven different stages of construction, both within the nucleus and after being exported into the cytoplasm.[26] In many organisms, mitochondria (i.e., organelles required for respiration in air) employ slightly variant genetic codes than the rest of the cell. This list could be expanded indefinitely. Suffice it to say that the complications of the machinery out of which organisms are built are nothing short of incredible, and these complications bear the stigmata of a complex history. Systems of control that turn on the machinery for making eyes in mice have recently been shown to turn on the machinery for making eyes in fruit flies.[27] Yet, the types of eyes that are made and the steps by which they are made are entirely different. The only sensible understanding of this mixture of conserved control systems and novel constructional apparatus is that the control system is a historical remnant, one that has – for reasons we do not understand very well – been retained for more than 500 million years. There can be no pretense of perfect adaptation here. Organisms are cobbled together. To use Jacob's word, they are constructed by a process of "bricolage" (roughly, tinkering), which means that the different parts of organisms and even cells (cf. mitochondria) have different histories (Jacob 1977, 1982). Careful examination of the current features of organisms thus reveals a great deal about the contingent pathways by means of which they were constructed. The importance of this deep truth within biology was central to Dobzhansky's perspective. To this extent, the last two decades of work in molecular biology have helped to justify his claim: Nothing in biology makes sense except in the light of evolution.

## REFERENCES

Adams, M. B. 1991. "Through the looking glass: The evolution of Soviet Darwinism." In *New Perspectives on Evolution*, eds. L. Warren and H. Kropowski. New York: Wiley-Liss, 37–63.

Adams, M. B., ed. 1994. *The Evolution of Theodosius Dobzhansky: Essays on His Life and Thought in Russia and America*. Princeton: Princeton University Press.

Allen, G. E. 1979. "Naturalists and experimentalists: The genotype and the phenotype." *Studies in the History of Biology* 3: 179–209.

Ayala, F. J. 1990. "Dobzhansky, Theodosius." In *Dictionary of Scientific Biography*, ed. F. L. Holmes. New York: Charles Scribner's Sons, 233–42.

[26] For amplification of these topics, see Chapters 9 and 11; for a stronger account of the nature and impact of recent findings, see Chapter 12.

[27] Halder, Callaerts and Gehring, 1995. Chapter 11 was an early reaction to the findings of the Gehring laboratory.

Beatty, J. 1991. "Genetics in the atomic age: The Atomic Bomb Casualty Commission, 1947–1956." In *The American Expansion of Biology*, eds. K. R. Benson, J. Maienschein, and R. Rainger. New Brunswick, NJ: Rutgers University Press, 284–324.

Bowler, P. J. 1983. *The Eclipse of Darwinism.* Baltimore: Johns Hopkins University Press.

Bowler, P. J. 1988. *The Non-Darwinian Revolution.* Baltimore: Johns Hopkins University Press.

Bowler, P. J. 1989. *The Mendelian Revolution.* Baltimore: Johns Hopkins University Press.

Burian, R. M. 1988. "Challenges to the evolutionary synthesis." *Evolutionary Biology* 23: 247–69.

Burian, R. M. 1996. "Underappreciated pathways toward molecular genetics as illustrated by Jean Brachet's chemical embryology." In *The Philosophy and History of Molecular Biology: New Perspectives*, ed. S. Sarkar. Dordrecht, Holland: Kluwer, 67–85.

Dawkins, R. 1976. *The Selfish Gene*. Oxford: Oxford University Press.

Dawkins, R. 1982. *The Extended Phenotype: The Gene as the Unit of Selection*. Oxford: Oxford University Press.

de Chadarevian, S., and J.-P. Gaudillière. 1996. "The tools of the discipline: Biochemists and molecular biologists." *Journal of the History of Biology* 29, 3: 327–462.

Depew, D. J., and B. H. Weber. 1995. *Darwinism Evolving: Systems Dynamics and the Genealogy of Natural Selection*. Cambridge, MA: MIT Press.

Dietrich, M. R. 1994. "The origins of the neutral theory of molecular evolution." *Journal of the History of Biology* 27: 21–59.

Dobzhansky, T. 1937. *Genetics and the Origin of Species*. New York: Columbia University Press.

Dobzhansky, T. 1956. *The Biological Basis of Human Freedom*. New York: Columbia University Press.

Dobzhansky, T. 1963. *Mankind Evolving: The Evolution of the Human Species*. New Haven: Yale University Press.

Dobzhansky, T. 1964. *Heredity and the Nature of Man*. New York: Harcourt, Brace, Jovanovich.

Dobzhansky, T. 1967. *The Biology of Ultimate Concern*. New York: New American Library.

Dobzhansky, T. 1973a. "Nothing in biology makes any sense except in the light of evolution." *American Biology Teacher* 35: 125–9.

Dobzhansky, T. 1973b. *Genetic Diversity and Human Equality*. New York: Basic Books.

Dobzhansky, T., and L. C. Dunn. 1946. *Heredity, Race, and Society*. New York: New American Library.

Dobzhansky, T., E. Boesiger, and B. Wallace, completed by 1983. *Human Culture: A Moment in Evolution*. New York: Columbia University Press.

Dobzhansky, T., F. J. Ayala, G. L. Stebbins, and J. W. Valentine. 1977. *Evolution*. San Francisco: Freeman.

Fisher, R. A. 1930. *The Genetical Theory of Natural Selection*. Oxford: Oxford University Press.

Fleming, D. 1968. "Émigré physicists and the biological revolution." *Perspectives in American History* 2: 152–89.
Gayon, J. 1990. "Critics and criticisms of the modern synthesis: The viewpoint of a philosopher." *Evolutionary Biology* 24: 1–49.
Gayon, J. 1992. *Darwin et l'après Darwin: Une histoire de l'hypothèse de sélection dans la théorie de l'évolution*. Paris: Kimé.
Gayon, J. 1998. *Darwinism's Struggle for Survival: Heredity and the Hypothesis of Natural Selection*. Transl. M. Cobb. Cambridge: Cambridge University Press.
Gould, S. J. 1982. "Introduction [to the reissued first edition, 1937]," T. Dobzhansky. *Genetics and the Origin of Species*. New York: Columbia University Press, xvii–xli.
Gould, S. J. 1983a. "The hardening of the modern synthesis." In *Dimensions of Darwinism*, ed. M. Grene. Cambridge: Cambridge University Press, 71–93.
Gould, S. J. 1983b. *Hen's Teeth and Horse's Toes*. New York: Norton.
Grene, M., ed. 1983. *Dimensions of Darwinism: Themes and Counterthemes in Twentieth-Century Evolutionary Theory*. Cambridge: Cambridge University Press.
Grene, M. 1987. "Hierarchies in biology." *American Scientist* 75: 504–10.
Grene, M. 1990. "Is evolution at a crossroads?" *Evolutionary Biology* 24: 51–81.
Haldane, J. B. S. 1932. *The Causes of Evolution*. London: Longmans.
Halder, G., P. Callaerts, and W. J. Gehring. 1995. "Induction of ectopic eyes by targeted expression of the eyeless gene in *Drosophila*." *Science* 267: 1788–92.
Harrison, R. G. 1937. "Embryology and its relations." *Science* 85: 369–74.
Hubby, J. L., and R. C. Lewontin. 1966. "A molecular approach to the study of genic heterozygosity in natural populations. I. The number of alleles at different loci in *Drosophila pseudoobscura*." *Genetics* 54: 577–94.
Jacob, F. 1977. "Evolution and tinkering." *Science* 196: 1161–6.
Jacob, F. 1982. *The Possible and the Actual*. Seattle: University of Washington Press.
Judson, H. F. 1996 (1979). *The Eighth Day of Creation: Makers of the Revolution in Biology*. Cold Spring Harbor, NY: Cold Spring Harbor Laboratory Press.
Lewontin, R. C. 1974. *The Genetic Basis of Evolutionary Change*. New York: Columbia University Press.
Lewontin, R. C. 1993. "Th. Dobzhansky – A theoretician without tools." In *Genetics of Natural Populations: The Continuing Importance of Theodsius Dobzhansky*, ed. L. Levine. New York: Columbia University Press, 87–101.
Lewontin, R. C., and J. L. Hubby. 1966. "A molecular approach to the study of genic heterozygosity in natural populations. II. Amount of variation and degree of heterozygosity in natural populations of *Drosophila pseudoobscura*." *Genetics* 54: 595–609.
Lewontin, R. C., J. A. Moore, W. B. Provine, and B. Wallace, eds. 1981. *Dobzhansky's Genetics of Natural Populations*. New York: Columbia University Press.
Mayr, E. 1959. "Where are we?" *Cold Spring Harbor Symposia on Quantitative Biology* 24: 1–15.
Mayr, E. 1962. "Accident or design: The paradox of evolution." In *The Evolution of Living Organism: Darwin Centenary Symposium of the Royal Society of Victoria*. Melbourne: Melbourne University Press, 1–14.
Mayr, E. 1963. *Animal Species and Evolution*. Cambridge, MA: Harvard University Press.

Mayr, E., and W. B. Provine, eds. 1980. *The Evolutionary Synthesis: Perspectives on the Unification of Biology*. Cambridge, MA: Harvard University Press.
Morange, M. 1994. *Histoire de la biologie moléculaire*. Paris: Editions de la Découverte.
Morange, M. 1998. *A History of Molecular Biology*. Transl. M. Cobb. Cambridge, MA: Harvard University Press.
Muller, H. J. 1950. "Our Load of Mutations." *American Journal of Human Genetics* 2: 111–76.
Nordmann, A. 1994. "The evolutionary analysis: Apparent error, certified belief, and the defects of asymmetry." *Perspectives on Science* 2: 131–75.
Ospovat, D. 1981. *The Development of Darwin's Theory*. Cambridge: Cambridge University Press.
Provine, W. B. 1971. *The Origins of Theoretical Population Genetics*. Chicago: University of Chicago Press.
Provine, W. B. 1980. "Introduction [Genetics]." In *The Evolutionary Synthesis*, eds. E. Mayr and W. B. Provine. Cambridge, MA: Harvard University Press, 51–8.
Provine, W. B. 1981. "Origins of the genetics of natural populations series." In *Dobzhansky's Genetics of Natural Populations, I–XLIII*, eds. R. C. Lewontin, J. A. Moore, W. B. Provine, and B. Wallace. New York: Columbia University Press, 1–83.
Sapp, J. 1987. *Beyond the Gene: Cytoplasmic Inheritance and the Struggle for Authority in Genetics*. New York: Oxford University Press.
Smocovitis, V. B. 1996. *Unifying Biology: The Evolutionary Synthesis and Evolutionary Biology*. Princeton: Princeton University Press.
Waddington, C. H. 1942. "Canalization of development and the inheritance of acquired characters." *Nature* 150: 563–5.
Waddington, C. H. 1960. "Experiments on Canalizing Selection." *Genetical Research* 1: 140–50.
Williams, G. C. 1966. *Adaptation and Natural Selection*. Princeton: Princeton University Press.
Wright, S. 1931. "Evolution in Mendelian Populations." *Genetics* 16: 97–159.
Wright, S. 1932. "The roles of mutation, inbreeding, crossbreeding, and selection." In *Proceedings of the Sixth International Congress of Genetics (Ithaca, NY)*. Menasha, WI: Brooklyn Botanic Garden, 356–66.
Zallen, D. T. 1993. "The 'light' organism for the job: Green algae and photosynthesis research." *Journal of the History of Biology* 26: 269–79.

# III

## Genetics and Molecular Biology

Shortly before moving to Virginia Tech in 1983, I began a study of gene concepts and how they changed over the years. Much of the work from this project has been published in article form, including three papers incorporated in this book as Chapters 7, 8, and 11. (For related papers, see Burian 1987, 1990, 2000; Burian and Zallen 2005 [in press]; Burian, Gayon, and Zallen 1988; Burian, Richardson, and Van der Steen 1996; Zallen and Burian 1992.)

Chapter 7 derives from a paper delivered at a 1984 conference on "The New Biology and the New Philosophy of Science," organized by David Depew and Bruce Weber at California State University, Fullerton. The main point of the chapter is that the classical gene concept is categorially open. By this, I mean that Mendelian gene concepts employed devices of *indefinite reference* that did not even specify the category to which genes belong – that is, if there really are such things as genes, what kind (or category) of entity they are. Thus, it was possible to identify and reidentify particular genes without knowing their exact constitution. I argue specifically that even though Bateson, one of the founders of genetics, thought that genes could not be "mere" material particles (he thought that they might be stable harmonic resonances of some sort),[1] whereas the Morgan group thought genes to be chemical units or material particles contained on chromosomes, Bateson and the Morgan group could agree on clear-cut criteria sufficient to determine when genes had been properly identified. This illustrates the importance of devices of indefinite reference – reference to a specific entity of unknown constitution that is inherited in a particular pattern, at least approximately, and is causally responsible for, say, the presence of vermilion (rather than red)

[1] See, for example, Chapter 2 of Bateson (1913). See also Chapter 8 of this book and the references in (Burian 2000), and for a new interpretation of Bateson, Newman (in press).

eyes in *Drosophila*. The chapter balances the philosophical work required to understand the referential devices required for such concepts to work with biological issues that go into understanding the classical gene concept.

Chapter 8 is descended from a paper delivered in a Mellon Workshop on "Building Molecular Biology" held at MIT in April 1992. It describes differences between biochemical and genetic approaches to the problem of protein synthesis in the early days of molecular biology. More specifically, it focuses on the role of problem orientation in shaping two distinct investigative pathways – namely, that of Paul Zamecnik, whose laboratory worked on biochemical analysis of the machinery for protein synthesis, and that of Jacques Monod, whose laboratory worked on biochemical and genetic analysis of the induction and regulation of protein synthesis.[2] Examination of the differences between these programs for studying protein synthesis opens up some of the difficult issues surrounding the impact of "molecularization" in reshaping biological research and on the disciplinary differences affecting biological knowledge, a theme that is taken up again in Chapters 9, 11, and 12.

Zamecnik's and Monod's programs of research yielded different standards for evaluating claims about the molecules that turned up in their research. This is not explicitly discussed in Chapter 8, which covers a period up to about 1960, but it is worth emphasizing here. Because the research programs were directed to different questions, they operated at different levels of abstraction. By 1961, Jacob and Monod spelled out their model of the operon (i.e., a switch that activated or deactivated the transcription of a group of genes). They argued that they had established the existence of a molecule participating in a specific biochemical process – namely, a "repressor." This molecule was not characterized biochemically and might, as far as the evidence showed, be a protein or an RNA. Its presence could be detected by the shutting down of an operon (Jacob and Monod 1961a, 1961b, 1962a, 1962b; Jacob, Gros, and Monod 1961; Monod and Jacob 1962; Monod, Jacob, and Gros 1961; Monod, Jacob, and Sussman 1962). This stand was quite foreign to that of the Zamecnik laboratory, which employed classical biochemical standards

[2] It is important not to import subsequent knowledge into the uncertain knowledge situation of the late 1950s. Until sometime in 1958, for example, Monod thought that even if amino-acid sequences were somehow determined by RNA formed from DNA, proteins were not synthesized *de novo*, but were made when some sort of cytoplasmic pre-enzyme, whose form or quantity was indirectly controlled by the genes, supposedly reshaped or reconfigured an indifferent substrate. See, for example, Monod (1958, p. 571), which claims that in forming ß-galactosidase "pre-enzyme combines specifically and reversibly" with a substrate, thus folding and shaping the substrate, which could form multiple globular enzyme proteins. Gaudillière (1991, pp. 57–9) provides more details on this position. See also Gaudillière (1992, 1993).

according to which – in order to have a convincing case that a molecule exists and takes part in a particular reaction – one must isolate the molecule, characterize it biochemically, and show directly that it enters into the reaction in question. Chapter 8 does not take the case study far enough forward to exhibit this difference in standards, but that difference is closely related to the differences between the problems that the two laboratories were addressing, discussed in the chapter. A laboratory aiming to elucidate the mechanism by which proteins are synthesized must provide full molecular detail. A laboratory aiming to show how protein synthesis is regulated may, in appropriate circumstances, be satisfied with showing that a determinately located regulatory switch is controlled by a chemical even though no one has determined the composition of the regulatory chemical or the mechanism by which the switch is thrown. In the experimental context, the power of the logical requirement for an entity satisfying such an indefinite description illustrates the influence of the genetic approach characterized in Chapter 7 and contrasts with the molecularly definite characterizations required by traditional biochemistry.

Chapter 9 was originally presented in January 1995 at a conference on gene concepts and evolution at the Max Planck Institute for the History of Science in Berlin and was previously published only in the form of a preprint (Burian 1995). Since that date, the Institute held a follow-up conference, yielding an outstanding volume that should be required reading for anyone who wishes to follow up on the topics covered here (Beurton, Falk, and Rheinberger 2000). Chapter 9 elaborates on the growing difficulty involved in understanding what discourse about genes refers to in light of the complexities uncovered by the molecularization of genetics, a topic pursued in greater depth in Chapters 11 and 12. On my account, molecular genes, like Mendelian genes, continue to be delimited by reference to the functions or phenotypes associated with effects that, in context, are ascribed to the presence or action of those genes. This mode of identification yields important constraints on the ways in which gene concepts refer – and preserve continuity of reference through conceptual change. It also constrains an account of what genes can turn out to be; they will have to be connected, however minimally, in appropriate ways to the phenotypes or functions of interest.

It will strike some readers as odd that what counts as a gene depends on the phenotypes of concern. But I argue that this is correct and that it helps make sense of how we retain the referential continuity of gene discourse as we gain more detailed understanding of the composition and behavior of the genetic material. Because the genetic material is not structured so that it has natural divisions corresponding to the functions or phenotypes of interest to us, there are no satisfactory means of providing strictly intrinsic characterizations of genes. This is a highly controversial claim, so I have added a postscript to this chapter to connect it to some developments in the recent literature on the topic of gene concepts.

## REFERENCES


Bateson, W. 1913. *Problems of Genetics*. New Haven: Yale University Press.

Beurton, P., R. Falk, and H.-J. Rheinberger, eds. 2000. *The Concept of the Gene in Development and Evolution: Historical and Epistemological Perspectives*. Cambridge and New York: Cambridge University Press.

Burian, R. M. 1987. "Realist methodology in contemporary genetics." In *The Process of Science*, ed. N. Nersessian. Dordrecht, Holland: Nijhoff, 195–210.

Burian, R. M. 1990. "La contribution française aux instruments de recherche dans le domaine de la génétique moléculaire." In *Histoire de la génétique: Pratiques, techniques et théories*, eds. J.-L. Fischer and W. H. Schneider. Paris: ARPEM, 247–69.

Burian, R. M. 1995. "Too many kinds of genes? Some problems posed by discontinuities in gene concepts and the continuity of the genetic material." In *Preprint 18: Workshop: Gene Concepts and Evolution – 6–7 January 1995*. Berlin: Max-Planck-Institut für Wissenschaftsgeschichte, 43–51.

Burian, R. M. 2000. "On the internal dynamics of Mendelian genetics." *Comptes rendus de l'Académie des Sciences, Paris. Série III, Sciences de la Vie/Life Sciences* 323, no. 12: 1127–37.

Burian, R. M., and D. T. Zallen. 2005 [in press]. "The gene." In *Cambridge History of the Life and Earth Sciences*, eds. P. J. Bowler and J. V. Pickstone. Cambridge: Cambridge University Press.

Burian, R. M., J. Gayon, and D. T. Zallen. 1988. "The singular fate of genetics in the history of French biology, 1900–1940." *Journal of the History of Biology* 21: 357–402.

Burian, R. M., R. C. Richardson, and W. J. Van der Steen. 1996. "Against generality: Meaning in genetics and philosophy." *Studies in the History and Philosophy of Science* 27: 1–29.

Gaudillière, J.-P. 1991. "Biologie moléculaire et biologistes dans les années soixante: La naissance d'une discipline. Le cas français." Université Paris VII.

Gaudillière, J.-P. 1992. "J. Monod, S. Spiegelman et l'adaptation enzymatique: Programmes de recherche, cultures locales et traditions disciplinaires." *History and Philosophy of the Life Sciences*, vol. 14, no. 1: 23–71.

Gaudillière, J.-P. 1993. "Molecular biology in the French tradition?" *Journal of the History of Biology* 26: 473–98.

Jacob, F., and J. L. Monod. 1961a. "Genetic regulatory mechanisms in the synthesis of proteins." *Journal of Molecular Biology* 3: 318–56.

Jacob, F., and J. L. Monod. 1961b. "On the regulation of gene activity: ß-galactosidase formation in *E. coli*." *Cold Spring Harbor Symposia on Quantitative Biology* 26: 193–211.

Jacob, F., and J. L. Monod. 1962a. "Sur la mode d'action des gènes et la régulation cellulaire." In *Semaine d'étude sur le problème des Macromolécules*: Académie Pontificale des Sciences, 85–95.

Jacob, F., and J. L. Monod. 1962b. "Déterminisme et régulation spécifique de la synthèse des protéines." In *Proceedings of the Fifth International Congress of Biochemistry*, 132–54.
Jacob, F., F. Gros, and J. L. Monod. 1961. "Sur la régulation et sur le mode d'action des gènes." *Journal de Chimie Physique* 58: 1100–2.
Monod, J. L. 1958. "An Outline of Enzyme Induction." *Recueil de Travaux en Chimie, Pays-bas et Belgique* 77: 569–85.
Monod, J. L., and F. Jacob. 1962. "Elements of Regulatory Circuits in Bacteria." In *Biological Regulation at the Cellular and Supercellular Level*. New York: Academic Press, 1–24.
Monod, J. L., F. Jacob, and F. Gros. 1961. "Structural and rate determining factors in the biosynthesis of adaptive enzymes." *Biochemical Society Symposia* 21: 104–32.
Monod, J. L., F. Jacob, and R. Sussman. 1962. "Sur la nature du répresseur assurant l'immunité des bactéries lysogènes." *Comptes rendus hebdomadaire de l'Académie des Sciences, Paris* 254: 4214–16.
Newman, S. E. In press. "Evo Devo 1.0: William Bateson's physicalist ideas." In *From Embryology to Evo-Devo*, eds. J. Maienschein and M. D. Laubichler. Cambridge, MA: MIT Press.
Zallen, D. T., and R. M. Burian. 1992. "On the beginnings of somatic cell hybridization: Boris Ephrussi and chromosome transplantation." *Genetics* 132: 1–8.

# 7

## On Conceptual Change in Biology

### The Case of the Gene[1]

The current situation in philosophy of science generally – and in philosophy of biology in particular – is most unsatisfactory. There are at least three general problems that many philosophers thought themselves near to solving in the 1960s–70s, only to find that the anticipated solutions have come unglued: (1) the problem of characterizing and understanding the dynamics of conceptual change in science; (2) the problem of understanding the relationships among theories (including in particular the reduction of one theory to another); and (3) the problem of scientific realism (i.e., the problem of how seriously to take the claims of theoretical science or, at least, of some theoretical scientists, to be describing the world literally in terms of such theoretical entities as genes, protons, DNA molecules, and quarks). This general situation has significant effects on the philosophical study of particular sciences. In philosophy of biology, for example, although one finds several elegant studies of particular topics, the sad fact is that there is no generally satisfactory large-scale synthesis in sight. We have no agreed-on foundation, no generally acceptable starting point from which to delimit and resolve the full range of theoretical problems of interest to scientists and philosophers regarding biology.

This chapter provides a preliminary report on a new approach to conceptual change, together with a sketch of its application to important biological subject matter. The approach offers some promise of providing a satisfactory framework, compatible with scientific realism, for detailed studies of particular scientific developments.

[1] Originally published in Depew and Weber (eds.), *Evolution at a Crossroads: The New Biology and the New Philosophy of Science*. Cambridge, MA: MIT Press, 1985, pp. 21–42. Reprinted by kind permission of MIT Press. Previous versions of this chapter were read at The University of California, Davis; California State University, Fullerton; and Virginia Polytechnic Institute and State University. Comments on these occasions and by friends too numerous too mention have greatly improved the paper. I am grateful to all concerned.

Before I sketch in some of the relevant philosophical background, it is useful to indicate how various concepts of the gene will enter the discussion. Mendel's work was "rediscovered" in 1900. Within fifteen years of that event – say, with the publication of *The Mechanism of Mendelian Heredity* by T. H. Morgan and his coworkers in 1915 (Morgan et al. 1915) – the main elements of the classical theory of the gene were fairly well established. A nearly constant series of improvements and refinements in that theory, grounded in good part in laborious but fascinating experimental work, resulted in considerable revision of the theory. Indeed, the accumulated changes run so deep that some have characterized this historical process as one in which Mendelian genetics was replaced by a series of improved successors that can be grouped under the label "transmission genetics." (For example, cf. Hull 1974, Chapter 1.) This extended process, both in its theoretical and empirical aspects, helped prepare the way for what is usually considered a scientific revolution brought on by the advent of molecular genetics. For present purposes, we may mark that advent by the publication of the justly famous solution of the principal structure of DNA in Watson and Crick (1953).

It is not possible to review the relevant history in this chapter – the task is simply too large.[2] Instead, I single out a couple of moments from the history of genetics to illustrate the character of conceptual change in that discipline and to support the claim that the conceptual changes examined fit well with the larger philosophical views put forward in this chapter. I show that the approach that biologists took to the conceptual changes in question had a significant effect on their practice, a result that suggests that the proper handling of conceptual change ought to be of real concern to working scientists. Finally, I suggest that a full-scale study of the concept of the gene is a singularly appropriate vehicle for working out a general account of conceptual change in science.

## CONTINUITY AND DISCONTINUITY IN SCIENTIFIC THEORIES

A crude description of the present impasse in the theory of conceptual change will suffice for present purposes.[3] Despite an immense variety of refinements, there are two main views to be considered. One of these, associated with Feyerabend (1962, 1965, 1970, 1975) and (perhaps mistakenly) Kuhn (1962,

[2] The best source for this history is Carlson (1966), which in part inspired this essay. Cf. also Whitehouse (1965) for a complementary approach.

[3] A useful elementary survey of the background is Brown (1977); a more advanced and detailed survey is Suppe (1977).

1970, 1977), might be labeled the discontinuity view. It claims that there are, at least occasionally, genuine conceptual revolutions in science. Revolutions are to be understood in a radical way; when they occur, those scientists who are separated by a revolution in a given field end up working with concepts and theories that are mutually incommensurable. This is to say that the pre- and post-revolutionary theories and concepts do not share a common denominator and, therefore, are not interdefinable in any useful way. Discontinuity theorists usually add that the recording of scientific observations requires some sort of conceptual apparatus and that the concepts involved in gathering and reporting scientific observations are theory-laden in some way or other. With these additions, they argue, it follows that observations worked up to support or test one of the theories in question need not (and in difficult cases will not) serve in any direct way to support, test, or undermine the competing theory. Otherwise, the relevant observational concepts could be used to provide at least partial interdefinitions of the concepts drawn from the competing theories; however, those theories were supposed, by hypothesis, to be incommensurable.

This result is counterintuitive. At first glance, at least, it seems simply incorrect to claim that evidence gathered under Newtonian or Mendelian auspices must automatically be thrown out or recast if it is to have a bearing on quantum mechanics or molecular genetics, respectively. Not surprisingly, such claims have been highly controversial, with some thinkers supporting them and others arguing that they reduce discontinuity views to absurdity. Indeed, the best-known theorist of scientific revolutions, Thomas Kuhn, backpedaled a long way from this kind of reading of his work in an attempt to avoid the "absurdities" that result from extreme interpretations of incommensurability (Kuhn 1983b). But, such extreme interpretations are not easily avoided. The difficulty is that people (including Kuhn who, as is shown in note 4, even now cannot escape the central difficulty) arrived at discontinuity theories honestly, by means of a series of persuasive arguments that have not been adequately answered.

Discontinuity theories arise from a set of seemingly commonsensical commitments regarding the nature of scientific language, concepts, and theories. Among these is the notion that a scientific theory is adequately represented as an interconnected body of statements about some domain of phenomena, and that this body of statements typically employs characteristic theoretical concepts. Theoretical concepts, in turn, cannot be properly understood in abstraction from the relevant theory. Although it may, arguably, be true that one need not employ the whole of a theory to understand or delimit the relevant concepts, one does presuppose at least some central core of the theory in question when utilizing those concepts. Thus, to measure the mass of a body, one must presuppose at least Newton's second law (F = ma) and probably the third law as well. And, for "mass" to refer (i.e., for bodies to genuinely have mass), bodies must in fact behave in accordance with Newton's laws; in short, the fundamental laws of Newtonian mechanics must be true if their theoretical terms are to refer to objects or properties in the world.[4] Similarly, to determine whether an organism is heterozygous for a recessive Mendelian gene – say, a gene for albinism – a classical geneticist must presuppose various theoretical principles. One principle claims, for example, that – other things being equal – an organism that has a recessive gene in double dose will, in the right circumstances, exhibit the trait (in our example, albinism) in terms of which the gene is identified. Obviously, on any reasonable account of the structure of classical genetics, further principles concerning the transmission of genes from parents to offspring must also be presupposed in setting forth the workings of heredity. And, organisms will possess genes only if the relevant Mendelian laws are true (or, perhaps, only if some fundamental subset of those laws is true).

These considerations show that a holistic account of theoretical language has important consequences. According to a holistic account, the proper employment of a theoretical concept requires one to mobilize some fairly large entity – an entire theory or paradigm, conceptual scheme or conceptual framework, or theoretical language or worldview, to borrow some of the terms current in the literature. Such holism brings radical incommensurability with it. What has led (or sometimes forced) people to swallow discontinuity theories has been their acceptance of some form of holism about theoretical concepts. They have been forced to this holism, in turn, by recourse to wholly inadequate semantic theories for theoretical languages.

The reason for claiming that Kuhn, like the rest of us, has not solved this tangle of problems is that he has not shown how to integrate his account of conceptual change into a nonholistic theory of theoretical language. What is required is some way to spring concepts like "mass" and "gene" sufficiently loose from the fundamental principles of the theories with which they are allied that it becomes apparent how one could seriously use those terms as a scientist without presupposing the truth of the corresponding principles. Only then will we be able to make sense of debates between theorists over which of their theories, built on incompatible principles, is a (or the) correct theory of mass or of the gene. It is not enough to wish to reject extreme

[4] Kuhn's original articulation of these ideas (1962, pp. 101–2) has been altered in various ways, but as Kuhn 1983a (pp. 566–7) shows, he still accepts the central claim that if the fundamental laws or postulates of a theory are not true, the leading theoretical terms of that theory do not refer. See the following note for an elaboration of this point.

versions of incommensurability; until philosophers show how to get around holistic theories of the language of theoretical science, they will be vulnerable to arguments forcing them into acceptance of radical incommensurability.[5]

For the sake of completeness, I briefly discuss the second class of theories of conceptual change in science – namely, continuity theories. At the moment, most extant continuity theories are discredited, for they are committed to the idea that some core of scientific concepts (e.g., so-called observational concepts) provides a permanent (although perhaps expandable) base for the spinning out of new theoretical concepts and theories. One reason for the disrepute into which continuity theories have fallen is their close connection to traditional philosophical accounts of the reduction of one theory to another. These accounts, alas, are a total failure. They claim that when a theory (say, Mendelian genetics) is reduced to (or by) another (say, molecular genetics), one can deduce the central claims of the theory that is reduced from those of the new, more fundamental theory.[6] The deduction uses the principles of the fundamental theory together with suitable definitions and statements connecting the concepts of the two theories and descriptions of the circumstances in which the reduced theory obtains. Were this account of reduction correct, the concepts of the reduced theory would be, in effect, definable within the more fundamental theory, and the claims of the reduced theory would be a subclass of the claims of the fundamental theory. In fact, as is now generally recognized, reduction of this sort virtually never occurs

[5] Kuhn's latest attempt to escape the consequences of holism turns on restricting the interconnections among terms and concepts to a local context – hence, his term "local holism." Yet, as the following quotation shows, Kuhn has serious trouble in accounting for disagreements over the reference of theoretical terms: "Nevertheless, I take the Second Law to be necessary in the following language-relative sense: If the law fails, the Newtonian terms in its statement are shown not to refer" (Kuhn 1983b, p. 567).

If this text is taken literally, Kuhn cannot properly parse a debate between an Einsteinian and a Newtonian physicist in which each individual argues that the theory he prefers gives a proper description of what, allegedly, they both refer to by the term "mass." This difficulty is a consequence of Kuhn's claim that if the Einsteinian term "mass" is well grounded, then the phonetically and lexicographically identical Newtonian term cannot refer at all. It follows that one cannot coherently argue that what is referred to by the Newtonian term is properly described within Einsteinian mechanics for one cannot coherently hold that the Newtonian term refers at all while also accepting Einstein's mechanics. Yet, such disagreements are absolutely commonplace and give no sign of incoherence. What has gone wrong is the mistaken linkage of reference to meaning. Even if holism is correct about the meanings of theoretical terms (as it surely is), such holism is mistaken when applied to reference. This point undercuts all standard treatments of incommensurability, including the weakened versions that Kuhn currently advocates while seeking to escape the consequences of radical versions of incommensurability.

[6] Cf., e.g., Nagel (1961). The extended debate among Hull 1974, 1976; Ruse 1976; Schaffner 1969, 1976; Wimsatt 1976, 1979; and others shows the sorts of difficulties that this approach encounters.

in science. One cannot even deduce Kepler's laws for our solar system from Newtonian mechanics. Given that there is more than one planet, what one deduces is the claim that the orbital formulae obtained from Kepler's laws are approximately correct although literally false. The calculable perturbations of Mars from its supposed elliptical orbit, for example, were observable within the limits of accuracy with which Kepler was working. Worse yet, when one comes to cases like statistical versus phenomenological thermodynamics or, as we will see, Mendelian versus molecular genetics, discontinuity theorists have put forth quite convincing arguments to show that the concepts of the theory to be reduced simply cannot be reproduced within the successor theory (Hull 1974, Chapter 1). This claim, as I argue, seems to be entirely in accord with the facts.

Over and above the connection between continuity theories and discredited theories of reduction, there is fairly broad consensus that they are not true to the facts. There simply is no timeless and secure set of observational concepts, let alone nonobservational concepts, strong enough to be employed as the base for building up the concepts of theoretical science.

How do all these abstract considerations bear on the workings of real science? What on earth is to be accomplished by dismissing both continuity and discontinuity theories of conceptual change? For those who wish to surmount difficulties, not just to find them, the next steps may provide some relief. I argue that the standard theories of conceptual change are vitiated by a mistaken presupposition that prevents their having real contact with live science. An examination of the concept – or, rather, the concepts – of the gene provides an ideal vehicle for bringing about a new start on a much more satisfactory footing.

## THE GENE: DEVELOPMENT OF A CONCEPT

This is time to make a first pass at various concepts of the gene. The question that we pursue is this: To what, if anything, do scientists refer to when they use the term "gene" or one of its cognates? For present purposes, this question provides enough of a guiding thread to lead us through what will someday have to be turned into an immensely complex discussion.

Honesty compels me to remind the reader how complex a full treatment of the history of the gene has to be. A single sentence from Carlson (1966) makes the point:

> The gene has been considered to be an undefined unit, a unit-character, unit-factor, a factor, an abstract point on a recombination map, a three-dimensional

> segment of an anaphase chromosome, a linear segment of an interphase chromosome, a sac of genomeres, a series of linear sub-genes, a spherical unit defined by target theory, a dynamic functional quantity of one specific unit, a pseudoallele, a specific chromosome segment subject to position effect, a rearrangement within a continuous chromosome molecule, a cistron within which fine structure can be demonstrated, and a linear segment of nucleic acid specifying structural or regulatory product (p. 259).

To cut through all this complexity, I make four fundamental points by use of one rather simple example. The first is that it is possible for scientists to exercise strong controls to ensure that they are referring to the same entity or entities despite large differences in viewpoint, terminology, concepts, and theoretical commitments. The episode in terms of which I make this point concerns William Bateson, one of the founding fathers of genetics – the one, in fact, who coined the term "genetics."[7] Bateson never was fully persuaded that genes could be localized on chromosomes or that they could be mere molecules or material particles. (Because of the dynamics required for them to achieve their effects, he thought that they would have to be stable harmonic resonances or something of the sort.) He usually employed the term "factor" rather than the later coinage "gene," and he used "character" or "unit character" for traits that he counted as the effects of the presence, absence, or (rarely) alteration of a single factor. He preferred to talk of "discontinuous variation" rather than "mutation," and he distinguished sharply between such discontinuous variations (which he thought were the real stuff of evolution) and the small continuous variations, which he took to be the target of Darwinian natural selection.

Bateson (1916) published a review of the Morgan group's definitive book, *The Mechanism of Mendelian Heredity* (Morgan et al. 1915), in *Science*.[8] Now, Bateson and Punnett had discovered linkage between genes (which they called "gametic coupling") in 1906. Ironically, linkage (which, by the way, requires violation of Mendel's law of independent assortment) served as one of the cornerstones for the Morgan group's arguments that chromosomes are the bearers of the hereditary material and the relative locations of the factors or genes may be pinpointed quite precisely by means of maps based on the degree of linkage between them. The point about the Bateson review is simple: Despite all the differences between his views and those of Morgan and his colleagues, there simply is no difficulty about the reference

[7] Useful secondary sources on Bateson include Cock (1983), Coleman (1970), and Darden (1977). For the invention of the term "genetics," see Carlson (1966, pp. 15–16).

[8] Cock (1983), who places considerably less emphasis than I do on Bateson's vibratory theory of heredity, has an extended discussion of Bateson's review of Morgan et al. at pp. 41ff.

of the relevant terms. Without explicit consideration, Bateson considered it to be established that certain traits (e.g., certain eye color traits in *Drosophila* that had been shown in experimentally controlled matings to be transmitted to offspring in well-established particulate patterns) are indeed the consequence of the presence (or perhaps, sometimes, the absence) of a particular factor. The main disagreements in question concern not whether there are factors or genes acting in such cases but rather what kinds of things these factors are and how strongly the available evidence supports the view that they are material particles (or some such) located at the relative positions on the chromosomes worked out by the Morgan group's mapping techniques. In the much later operationalist terminology of Stadler (1954), Bateson retreated to an operational definition of the gene, which he shared with Morgan's group; his disagreement with them concerns the best theoretical account to offer of the behavior of the operationally defined gene. In Stadler's terminology, this disagreement concerns the hypothetical gene. (I return briefly to Stadler's distinctions at the end of this chapter.)

The second point to be drawn from this episode is that the procedures involved cannot be confined to Bateson's and Morgan's groups; they are already the firmly established property of a larger community. It is, of course, true that the procedures by which factors (or genes) are identified and individuated can be refined and improved in ways that may shift the reference of particular terms; such refinements can change altogether the set of entities to which a community refers by use of such terms as "gene" and "factor." But, the terms are – in a certain sense – community property. There are experts who are in a position to adjudicate questions regarding whether Morgan and his colleagues met the conditions required for them to be referring – at least, *prima facie* – to particular genes and to determine whether any objections to their claim so to have referred are cogent or not. Bateson, himself one of those experts, knew that the Morgan group had unassailably demonstrated that they were dealing with single genes by the standards then available.

This point about the social character of the referential use of scientific and prescientific terms is rather stronger than it looks. As a concrete example which, in spirit, goes back to the work of Putnam (1975, pp. 223–9 ff.), I am able to use the terms "mole" and "mole rat" to refer to distinct mammals despite the fact that if someone were to bring a few of each of them to me I could not tell which were the moles and which the mole rats. The reason that my lamentable ignorance does not prevent me from using these terms to refer correctly to distinct animals is that my usage is interlocked quite deliberately with that of a larger community – particularly with that of experts, to whom I am prepared to defer, who can tell moles from mole rats.

The matter is, of course, not so simple when it comes to questions on the forefront of knowledge; all of the experts may be (and have on certain issues been) wrong in their central beliefs about genes. Nonetheless, as the example of the Bateson review demonstrates, when the social controls work correctly and the world cooperates (neither of which can be counted on), the experts' procedures allow them to secure reference for theoretical terms like "factor" and "gene" even in the face of considerable disagreement, even in the face of pervasive and fundamental error in the foundations of their theories.

The third point suggested by consideration of the Bateson review is that terms like those under discussion are used by the community in a way that does not require that their reference be fully specified. To put the point differently, the account we give of the reference of terms like "gene" ought to allow radically different descriptions of the things that genes might turn out to be. We often secure reference for theoretical terms and concepts while using deeply mistaken theories. It is clear, I hope, that if the world were appropriately different and if the genes in question in the Bateson review were, in fact, stable harmonic resonance rather than segments of chromosomes, nothing would need to be changed in what Bateson or the Morgan group wrote, but the reference of some of their terms would have been different. To use a term from Kitcher (1978, 1982), we must consider not only the reference but also the "reference potential"[9] of a term to understand how it is used. One can argue that, as of 1916, the terms used to refer to, say, the vermilion eye color gene of *Drosophila melanogaster* included stable harmonic resonances in their reference potential. Not long afterward, thanks to the Morgan group's success, that reference potential shifted; some genes, at least, might be localized on chromosomes. Accounting for such changes is an important part of a satisfactory history of the concept of the gene.

The fourth point is more or less a corollary of the third. It is sometimes possible to resolve disputes regarding the reference of theoretical terms in the light of subsequent investigation. No matter what Bateson thought he was referring to when he spoke of the *Drosophila* genes, he was, in fact, referring to segments of *Drosophila* chromosomes. Sometimes, at least, the world is well enough behaved to allow a clean resolution of such disputes. At the same time, to gain a proper appreciation of Bateson's writings, we must not let the truth of the matter run away with us. Bateson's beliefs inevitably affected his

[9] My encounter with this notion in Kitcher (1978) was crucially formative. This chapter grew out of thinking through how to apply the apparatus developed there to the history of the concept of the gene. His development of this notion in Kitcher (1982, pp. 339–47) is particularly useful. I am also grateful to Kitcher for helpful discussions concerning our work in progress; the overlap of interest and approach is unusually strong.

terminology; in this, he is typical of all working scientists. Sometimes he used the term "factor" in the attempt to refer to the entities that his own undeveloped theory, if correct, would have described. In such instances (because there are no such things), the term does not, in fact, refer at all. At other times, he used the term "factor" to refer to those entities – whatever they might turn out to be – that he, the Morgan group, and many others had singled out by their experiments. The unwary reader who fails to distinguish among these different uses of the same term will be unable to evaluate Bateson's claims correctly.

## THEORIES OF REFERENCE AND THE GENE

Let us collect our results to date. From a philosophical point of view, they concern the theory of reference and the proper analysis of theoretical concepts. I discuss each of these in turn.

At the moment, there are two fairly standard theories of reference with which to cope. The traditional theory (usefully described in Schwarz 1977), the roots of which lie in the work of Frege, holds that kind terms like "gene" refer, in the first instance, to any and all objects that fit the underlying description or descriptions with which the speaker is prepared to back up his or her use of the term. Those descriptions are said to spell out the sense of the term in question. According to the traditional theory, the reference of a kind term is fixed by its sense and by the world; a kind term refers to just those objects that fit the description or descriptions implicit in the sense of the term.[10]

In the context of the present investigation, this traditional theory can be seen to be allied with the holistic theories of theoretical language rejected previously. Its natural application to the Bateson case employs Bateson's mistaken and unarticulated theory of factors to determine what the sense of "factor" is and thus, in turn, to fix what he referred to in his primary uses of that term. The consequence is that the traditional theory claims that – precisely because Bateson's theory of factors (or genes) was fundamentally mistaken – Bateson did not refer to anything when he used whichever term for genes. On at least some occasions, in contrast, the Morgan group did refer to certain chromosome segments by use of such terms as "gene" and "factor." But, this contrast is wrong and wrongheaded. One small symptom of its erroneousness

[10] The power of this position is illustrated by Kuhn's tacit reliance on it in the argument criticized in note 4. If it were not for the supposed unique connection between sense and reference, what ground would there be for holding that – given the way the world is – the reference of a theoretical term or concept is fully fixed by its place in the fundamental laws of that theory?

is that, taken seriously, it does not allow one to construct a plausible construal of Bateson's review.[11]

Although there are a number of moves that one can make to try to save the traditional theory in application to this case, this is not the occasion to explore them. I simply assert, dogmatically, that the traditional theory is wrong and ask its defenders to present their arguments to the contrary.

The alternative theory, various versions of which have been developed by Donnellan, Kripke, Putnam, and others (cf., e.g., Donnellan 1966, 1970; Kripke 1972; Putnam 1973, 1975), is sometimes known as the "causal theory of reference." (The label is a misnomer, but the issues involved are not of immediate importance to the matters at hand.) At first sight, the causal theory fares rather better in dealing with our case. It holds that what a kind term like "gene" refers to depends on the earlier uses back to which present uses can be traced. "Gene" refers to whatever natural kind (assuming that there is one) in fact entered into the relevant causal interactions when Mendel performed his experiments on garden peas and baptized the factors that determined the patterns of inheritance that those peas exhibited. Thus, because in the tradition we are exploring, the term "gene" can be traced back to Mendel's uses of the German words *Charakter*, *Element*, *Faktor*, and *Merkmal* as means of describing the determinants of particulate inheritance, the causal theory says that both Bateson and Morgan were referring to the determinants of particulate inheritance, whatever their true nature.

*Prima facie*, this theory of reference can handle the Bateson review. As usually developed, however, like the Fregean theory, it is a closed rather than an open theory of reference.[12] By this, I mean that the reference of such kind terms as "gene" is treated as being completely fixed once those terms are properly established within the tradition. This closure, I believe, is yet another reflection of a holistic theory of theoretical language. Indeed, the fundamental presupposition shared by continuity and discontinuity theories of conceptual change is just this idea that the reference of theoretical concepts is closed. Closure of reference makes conceptual change into an all-or-nothing phenomenon.[13] When we turn to the history of genetics for the second time, we will see that not only the reference potential but also the actual reference

[11] The point is entirely parallel with the objection to Kuhn put forward in note 4.

[12] Kitcher makes a similar point in different terminology in Kitcher (1982, p. 345).

[13] Closure of reference with respect to theoretical terms is the fundamental move that undermines "local holism." So long as the fundamental laws of the relevant theory or the particular linguistic practices of a founding individual or community are thought to fix the reference of theoretical terms once and for all, theory change that adjusts the reference of the resulting theoretical terms will be impossible, and referential change regarding theoretical entities will be an all-or-nothing phenomenon.

of the term "gene" has changed in a controlled way during the development of the discipline of genetics.

But, before turning again to cases, a few words are needed about reference potential. Following Kitcher (1978, 1982), although with important modifications,[14] I maintain that reference potential provides the central tool by means of which to analyze theoretical concepts. In the Fregean tradition (see Frege 1892 (1952)), the concept mobilized by the use of a term is equated with the sense of the term – that is, with the underlying descriptions that speakers would employ to back up their use of the term. The alternative account that I am about to sketch is designed to reflect precisely those characteristics of the referential use of theoretical kind terms that give the Fregean tradition the greatest difficulty. Specifically, we need to consider the ways in which the practice of scientists makes three things likely: first, that their fundamental theoretical terms will pick out some natural kind even when their theories about that kind are badly mistaken; second, that different scientists will be able to refer to the same natural kinds even when their theories about those kinds are in radical disagreement; and, third, that the precise referential use of such central theoretical terms as "gene" can, in good cases, be brought into line with experimental results and with the theoretical commitments of the relevant community once that community achieves full consensus on those commitments.

## REFERRING TO GENES

The price that must be paid for theoretical discourse to have these characteristics is the referential openness of theoretical concepts and systematic ambiguity in the referential use of theoretical terms. Some will consider this price too high, but it has always been paid, for it is unavoidable. The history of genetics nicely illustrates the openness and systematic referential ambiguity of theoretical terms as well as some of the ways in which these are controlled. (The phenomena involved, however, are found in all of the sciences.)

Given the success of the Morgan group, the term "gene" typically refers, in fact, to a segment of a chromosome that, when activated or deactivated, performs a certain function or has a characteristic effect. But, how much of a chromosome? And what functions or effects? Much of the effort that went into mapping genes may be viewed as an attempt to answer the first question; much labor was expended on the determination of which part of

[14] Among the modifications: Kitcher seems to think that reference potential is an extensional concept. On my reading, it is not. Thus, he claims to be an extensionalist. I am not.

which chromosome contained which genes. In the process, certain criteria were developed for telling one gene from another. According to one of these, if two mutations affecting the same phenotypic trait – say, two eye-color mutations – could be separated by recombination, then they belonged to separate genes; if they could not be so separated, then they belonged to the same gene (i.e., they were counted as alternative alleles at the same genetic locus). This way of individuating genes was suggested by Sturtevant (1913a, 1913b), who suggested that two closely linked eye-color mutations (called "white" and "eosin") that Morgan and Bridges had been unable to separate in an experiment using 150,000 flies (Morgan and Bridges 1913) should be considered two alternative abnormal alleles at a single locus, the locus that had already been located in a specific position on the X chromosome. Now, the more closely two genes are linked, the more difficult it is to separate them by recombination and the larger the number of flies that must be used to execute the test. Thus, it should be no surprise that such claims are sometimes wrong and that it was established many years later that, in this very case, one can separate the two genes in question if one performs a truly gigantic recombination experiment (cf. Carlson 1966, p. 64; Kitcher 1982, p. 351).[15]

Consider the problem this creates when one asks what is referred to by subsequent uses of such terms as "the gene for white eyes" or "the eosin locus." If one conforms to the usage established on the basis of Sturtevant's results, one refers to that portion of the chromosome that contains both the white and the eosin genes. But, if one is working with the recombination criterion for theoretical purposes, one may refer instead to the smaller portion of the chromosome containing one but not both of these genes. This is to say that two rather different segments of the chromosome belong to the reference potential of these phrases. Often, it makes no difference which portion of the chromosome one refers to; they are, after all, virtually inseparable by ordinary techniques. But, occasionally, it may matter whether one purpose or the other – conformity to established usage in order to accomplish coreference with other scientists or correct application of the criteria separating genes from one another – dominates one's usage. For a long time, the ambiguity was inescapably built into the mode of reference that was available in discussing these genes.

Indeed, at various stages in the history of genetics, it became a theoretical and practical necessity to distinguish between different gene concepts, each of

[15] Cf. Carlson (1966, Chapter 8) for a discussion of the conceptual importance of Sturtevant's analysis, which provided the key step in recognizing that mutation often involves alteration rather than loss of genes.

which selected different segments of the chromosome or employed different criteria of identity for genes – remember Carlson's (1966) list! For example, in the 1950s, Benzer pointed out that many geneticists had assumed that the smallest unit of mutation with a distinct functional effect coincided with the smallest unit of recombination – and he performed some elegant experiments that showed that this claim is false (Benzer 1955, 1956, 1957). As a result, in some circumstances, it became necessary to choose among the unit of function (which, for reasons that need not concern us, Benzer called the "cistron"), the unit of mutation (which he called the "muton"), and the unit of recombination (which he called the "recon"). This particular result showed that there had been hidden openness in the reference potential of the term (and the concept) "gene" and that, in some arguments, although not in general, it was necessary to divide the reference of that term (concept) according to the separable modes of individuating genes.

The actual history is, of course, much richer than I have let on here, particularly when one pursues the story into the present, where one encounters split genes with separately movable subunits, transposable control elements, parasitic ("selfish") DNA, and so on. But, enough has been said to show that there are at least four ways in which the reference of a particular use of the term "gene," or one of its cognates, might be specified (compare Kitcher's prior discussion in Kitcher 1982, pp. 342 ff.). Which one of these is relevant will turn on the dominant intention of the scientist and the context of the discussion. One such intention is conformity to conventional usage. Taking Sturtevant's (1913a, 1913b) experimental result for granted, conformist usage would refer to the same segment of the X chromosome whether one spoke of the white or the eosin locus. Another sometimes conflicting intention is accuracy in the application of the extant criteria for identifying the relevant kinds or individuating the individuals of those kinds. When accuracy is the dominant intention, "white" and "eosin" refer to different segments of the chromosome. Thus, Sturtevant's mistake expanded the reference potential of the term "gene" by adding a compound chromosomal segment to the items potentially referred to by that term. In some (but only a few) contexts, it proved terribly important to consider the resultant long-unrecognized ambiguity of reference to understand the actual use of the relevant terms and to reconcile conflicts between competing descriptions of the experiment outcomes. What is at stake here is the precise roles that one's theoretical presuppositions and accepted experimental results play in fixing the reference of one's terms. Although this discussion has not provided a general resolution of that difficult problem, it has given some indication of the proper apparatus to employ in carrying out case-by-case analyses.

The Benzer case illustrates a third way in which reference may be fixed: Once an ambiguity (e.g., between "cistron" and "recon") becomes troublesome, it is sometimes necessary to stipulate as clearly as possible which of the available options one is taking as a way of specifying the reference of one's terms. Even at the risk of total failure to refer – which might happen if one's analysis is mistaken – one fixes one's reference to all and only those things that fit a certain theoretical description. The result is clarity and, when clarity is the dominant intention, reference is fixed in much the way that Frege thought that it is always fixed. A sense is determined by a description, and reference depends on whether anything, in fact, fits that description. Finally, one may operate with a dominant intention that Kitcher calls naturalism – namely, the intention to refer to the relevant effective natural kind occurring or operating in a certain situation or in a certain class of cases. Although the matter needs to be argued on another occasion, I suspect that one must have recourse to naturalism over and above conformity, accuracy, and clarity in order to put forth a successful account of the grounds on which Mendel, Bateson, Morgan, Benzer, and all the rest may be construed as employing concepts of the same thing: the gene.

## CONCLUDING REMARKS

To conclude this chapter, it is useful to comment briefly about the relationship between Mendelian and molecular genetics. As the reference of the term "gene" became more tightly specified during the development of Mendelian (or, if you prefer, transmission) genetics, in a large range of central cases the concept of the gene became that of a minimal chromosomal segment performing a certain function or causing a certain effect. The relevant effect was known as the phenotype of the gene. Not surprisingly, a major part of the history of the gene (not addressed here) concerns the interplay between what one counts as genes and how one restricts or identifies the phenotypes that can be used to specify individual genes. But, when all this is said and done, a great variety of phenotypes can legitimately be used to single out genes. As will soon be made clear, one's very concept of a gene depends on the range of phenotypes one considers.

Thanks to the advances made in molecular genetics, it is now possible to examine changes in the DNA (mutations!?) fairly directly. In some cases, at least, it is also possible to track the effects of those changes rather exactly. It is now well known that some changes in the DNA are silent; that is, they have no effect on any other aspect of the structure, the development, or the composition of the organism. Effectively, such changes in the genetic material do not amount to changes in the function of any gene; however, when suitably located, they do constitute changes in the structure or composition of the relevant gene. Other changes in the DNA do, of course, change the organism, but some of them do so in ways that, arguably, are of no importance to its structure, development, or function. For example, some so-called point mutations result in the substitution of one amino acid for another in some particular protein manufactured in accordance with the information contained on the gene in question. Many such substitutions have drastic effects. But some of them, so far as can be told, do not significantly alter the way the protein folds, its biological activity or its function in any significant way. In such cases, I suggest, there are strong reasons for tolerating in perpetuity important ambiguities in the concept of the gene.

The reason for this is that phenotypes at different levels are of concern for different purposes. Consider, for example, medical genetics. If one is concerned with PKU and allied metabolic disorders, the phenotypes one deals with will range all the way from gross morphological and behavioral traits to enzyme structure and function.[16] With respect to all of these phenotypes, both silent changes in the DNA and those that have no effect on enzyme structure or function will not (and should not) count as mutations (i.e., as changes in the relevant gene). It does not matter whether these changes occur within that segment of DNA that constitutes the gene of interest; because they have no relevant functional effects, the gene counts as unchanged. The reason for this is clear: The concept of the gene is coordinate with the concept of the phenotype. The phenotype of concern is not defined biochemically at the level of DNA but (if it is defined biochemically at all) rather at the level of protein or via some functional attributes consequent on the biochemistry of the relevant proteins.

It is important to recognize that there are legitimately different interests that lead us to deal with different types of phenotypes. Evolutionists, for example, may be interested in the rate of amino-acid substitutions in proteins or of nucleotide substitutions in DNA; that is, the phenotypes they are concerned with might be defined by amino-acid or even nucleotide sequence, not protein function. Accordingly, their definitions of the phenotype and the gene may be discordant with those of the medical geneticist. It is not a matter of right or wrong but simply a matter of legitimately different interests. The point is a fairly deep one, allied to Putnam's point discussed previously, about the linguistic division of labor. I have written, until now, sloppily, as if there were

[16] There is a helpful discussion of PKU in Burian (1981–2, pp. 55–9).

only one community of biologists or, rather, geneticists; this is simply false. There are large and important specialized subcommunities with legitimately different interests – interests that lead them to deal with legitimately different phenotypes. As the example just introduced shows, there are serious cases in which there is no question but that those differing phenotypes correspond with different concepts of the gene and different criteria for individuating genes.

Work in molecular genetics may well show that, like Bateson's, some contemporary attempts at establishing gene concepts are ill founded. Indeed, I believe that there are clear cases – for example, in sociobiology (cf. Burian 1981–2) – but perhaps much more generally, in which certain gene concepts will simply have to be abandoned in light of some of the findings of molecular genetics. But, molecular genetics cannot discriminate among well-founded gene concepts. There is a fact of the matter about the structure of DNA, but there is no single fact of the matter about what the gene is. Even though their concepts are discordant, the community of evolutionists concerned with the evolution of protein sequence and the community of medical geneticists working on metabolic disorders are both employing perfectly legitimate concepts of the gene. This provides strong, concrete support for the claim that the concept of the gene is open rather than closed with respect to both its reference potential and its reference.

A dangling thread provides a moral for biologists to consider. Recall Stadler's distinction between the operational concept of the gene and the various hypothetical concepts of the gene. Stadler is correct that proper use of an operational concept can ensure conformity and protect against the pernicious effects of certain theoretical errors. But, as the example of white and eosin genes shows, operational criteria (here, specifically for the individuation of genes) are themselves theory-laden and quite often erroneous. Furthermore, there is no single operational concept (or set of operational criteria) for the gene. In the end, as this brief discussion of molecular genetics suggests, the best arbiter we have of the legitimacy of both operational and hypothetical concepts of the gene comes from molecular analysis. The latter, in turn, cannot be extricated from what Stadler would have considered a hypothetical concept: namely, that of the structure of the DNA molecule. It follows that genetic concepts (and theoretical concepts generally) are inescapably open in the ways I have been describing.

For philosophical readers, I add only that the results of this extremely sketchy treatment of the case of the gene, if they withstand scrutiny, will prove to be of immense consequence for our understanding of reduction and of scientific realism.

REFERENCES


Bateson, W. 1916. "[Review of] Morgan, Sturtevant, Muller, and Bridges, *The Mechanism of Mendelian Heredity*." *Science* 44: 536–43.

Benzer, S. 1955. "Fine structure of a genetic region in bacteriophage." *Proceedings of the National Academy of Sciences, USA* 41: 344–54.

Benzer, S. 1956. "Genetic fine structure and its relation to the DNA molecule." *Brookhaven Symposia in Biology* 8: 3–16.

Benzer, S. 1957. "The elementary units of heredity." In *A Symposium on the Chemical Basis of Heredity*, eds. W. D. McElroy and B. Glass. Baltimore: Johns Hopkins University Press, 70–93.

Brown, H. I. 1977. *Perception, Theory, and Commitment: The New Philosophy of Science*. Chicago: Precedent and University of Chicago Press [1979].

Burian, R. M. 1981–2. "Human sociobiology and genetic determinism." *Philosophical Forum* 13: 40–66.

Carlson, E. A. 1966. *The Gene: A Critical History*. Philadelphia and London: W. B. Saunders.

Cock, A. G. 1983. "William Bateson's rejection and eventual acceptance of the chromosome theory." *Annals of Science* 40: 19–59.

Coleman, W. 1970. "Bateson and chromosomes: Conservative thought in science." *Centaurus* 15: 228–314.

Darden, L. 1977. "William Bateson and the promise of Mendelism." *Journal of the History of Biology* 10: 87–106.

Donnellan, K. 1966. "Reference and definite descriptions." *Philosophical Review* 75: 281–304.

Donnellan, K. 1970. "Proper names and identifying descriptions." *Synthese* 21: 335–58.

Feyerabend, P. 1962. "Explanation, reduction, and empiricism." In *Minnesota Studies in the Philosophy of Science*, eds. H. Feigl and G. Maxwell. Minneapolis: University of Minnesota Press, 28–97.

Feyerabend, P. 1965. "Problems of empiricism." In *Beyond the Edge of Certainty, University of Pittsburgh Series in the Philosophy of Science*, ed. R. G. Colodny. Englewood Cliffs, NJ: Prentice-Hall, 145–260.

Feyerabend, P. 1970. "Problems of empiricism, part II." In *The Nature and Function of Scientific Theories, University of Pittsburgh Series in the Philosophy of Science*, ed. R. G. Colodny. Pittsburgh: University of Pittsburgh Press, 275–353.

Feyerabend, P. 1975. *Against Method*. London: New Left Books.

Frege, G. 1892. "Über Sinn und Bedentung." *Zeitschrift für Philosophie und philosophische Kritik* C: 25–50. Translated as "On sense and reference" by Max Black. In *Translations from the Philosophical Writings of Gottlob Frege*. 1950, eds. P. Geach and M. Black. Oxford: Blackwell, pp. 56–77.

Hull, D. L. 1974. *Philosophy of Biological Science*. Englewood Cliffs, NJ: Prentice-Hall.

Hull, D. L. 1976. "Informal aspects of theory reduction." In *PSA 1974*, eds. R. S. Cohen, C. A. Hooker, A. C. Michalos, and J. W. V. Evra. Dordrecht, Holland: Reidel, 653–70.

Kitcher, P. 1978. "Theories, theorists and theoretical change." *Philosophical Review* 87: 519–47.

Kitcher, P. 1982. "Genes." *British Journal for the Philosophy of Science* 33: 337–59.

Kripke, S. 1972. "Naming and necessity." In *Semantics of Natural Language*, eds. D. Davidson and G. Harman. Dordrecht, Holland: Reidel, 253–355.

Kuhn, T. S. 1962. *The Structure of Scientific Revolutions*. Chicago: University of Chicago Press.

Kuhn, T. S. 1970. "Reflections on my critics." In *Criticism and the Growth of Knowledge*, eds. I. Lakatos and A. Musgrave. Cambridge: Cambridge University Press, 231–78.

Kuhn, T. S. 1977. *The Essential Tension*. Chicago: University of Chicago Press.

Kuhn, T. S. 1983a. "Commensurability, comparability, communicability." In *PSA 1982*, eds. P. D. Asquith and T. Nickles. East Lansing: Philosophy of Science Association, 669–88.

Kuhn, T. S. 1983b. "Rationality and theory choice." *Journal of Philosophy* 80: 563–70.

Morgan, T. H., and C. B. Bridges. 1913. "Dilution effects and bicolorism in certain eye colors of *Drosophila*." *Journal of Experimental Zoology* 15: 429–66.

Morgan, T. H., A. H. Sturtevant, H. J. Muller, and C. B. Bridges. 1915. *The Mechanism of Mendelian Heredity*. New York: Henry Holt and Co.

Nagel, E. 1961. *The Structure of Science*. New York: Harcourt, Brace, and World.

Putnam, H. 1973. "Explanation and reference." In *Conceptual Change*, eds. G. Pearce and P. Maynard. Dordrecht, Holland: Reidel, 199–221.

Putnam, H. 1975. "The meaning of meaning." In *Philosophical Papers*. Cambridge: Cambridge University Press, 215–71.

Ruse, M. E. 1976. "Reduction in genetics." In *PSA 1974*, eds. R. S. Cohen, C. A. Hooker, A. C. Michalos, and J. W. V. Evra. Dordrecht, Holland: Reidel, 633–51.

Schaffner, K. F. 1969. "The Watson–Crick Model and reductionism." *British Journal for the Philosophy of Science* 20: 325–48.

Schaffner, K. F. 1976. "Reductionism in biology." In *PSA 1974*, eds. R. S. Cohen, C. A. Hooker, A. C. Michalos, and J. W. V. Evra. Dordrecht, Holland: Reidel, 613–32.

Schwarz, S. P. 1977. "Introduction." In *Naming, Necessity, and Natural Kinds*, ed. S. P. Schwarz. Ithaca: Cornell University Press, 13–41.

Stadler, L. J. 1954. "The gene." *Science* 120: 811–19.

Sturtevant, A. H. 1913a. "A third group of linked genes in *Drosophila ampelophila*." *Science* 37: 990–2.

Sturtevant, A. H. 1913b. "The linear arrangement of six sex-linked factors in *Drosophila*, as shown by their mode of association." *Journal of Experimental Zoology* 14: 43–59.

Suppe, F. 1977. "Introduction" and "Afterword – 1977." In *The Structure of Scientific Theories, 2nd ed.* Urbana: University of Illinois Press, 3–232, 617–730.

Watson, J. D., and F. H. C. Crick. 1953. "Molecular structure of nucleic acids." *Nature* 171: 737–8.

Whitehouse, H. L. K. 1965. *Towards an Understanding of the Mechanism of Heredity*. Stanton, England: Arnold.

Wimsatt, W. C. 1976. "Reductive explanation: A functional account." *PSA 1974*: 671–710.

Wimsatt, W. C. 1979. "Reduction and reductionism." In *Current Problems in the Philosophy of Science*, eds. H. Kyburg, Jr., and P. Asquith. East Lansing: Philosophy of Science Association, 352–77.

# 8

# Technique, Task Definition, and the Transition from Genetics to Molecular Genetics

## Aspects of the Work on Protein Synthesis in the Laboratories of J. Monod and P. Zamecnik[1]

> In biology proteins are uniquely important. They are not to be classed with polysaccharides, for example, which by comparison play a very minor role. Their nearest rivals are the nucleic acids ... *The main function of proteins is to act as enzymes.*
>
> ... In the protein molecule Nature has devised a unique instrument in which an underlying simplicity is used to express great subtlety and versatility; it is impossible to see molecular biology in proper perspective until this peculiar combination of virtues has been clearly grasped (Crick 1958).

### INTRODUCTION

This epigraph, from Francis Crick's seminal article "On Protein Synthesis," can serve as a reminder that it is important to distinguish between molecular genetics and molecular biology. Scientists have often used the terms "molecular biology" and "molecular genetics" interchangeably, thus confounding the two. As Zallen (1996) has pointed out, however, this usage has always been problematic. I maintain that the failure to distinguish between molecular genetics and molecular biology bears on an important historiographic issue: the nature of scientific disciplines. On the account I advocate, it is important to counteract the confusion between the two

[1] Published in *Journal of the History of Biology* 26 (1993): 387–407. © 1993 Kluwer Academic Publishers. Reprinted with kind permission of Kluwer Academic Publishers. An early version of this chapter was written during a residential fellowship at the National Humanities Center, with financial support from the National Endowment for the Humanities and a Study/Research Leave from Virginia Polytechnic Institute and State University. The material and moral support of these sponsors is greatly appreciated. The paper was greatly improved by discussion with participants at the Mellon Workshop on Building Molecular Biology (MIT, April 1992), at a colloquium of the Center for the Study of Science in Society at Virginia Tech, and with Muriel Lederman.

because molecular genetics is clearly a discipline, whereas molecular biology is not.

On my rather traditional account, disciplines are organized and institutionalized bodies of research focused around a core group of questions. Molecular biology, taken widely, is extremely well organized and institutionalized; nonetheless, on my account it is not a discipline because it does not center on a focal group of questions. Molecular biology, after all, studies – among many other things – the structure and behavior of proteins but also of polysaccharides, lipids, lysosomes, ribosomes, membranes, muscle fibrils, and so forth. Molecular biology is thus a technique-based field that impinges on or includes a number of disciplines, many interdisciplinary investigations, and many investigations whose disciplinary location, if any, is uncertain. Molecular genetics, in contrast, is not only an organized and institutionalized domain of scientific research, it is also centered around a reasonably compact set of well-circumscribed focal questions.

This point is not merely terminological. Whatever words we employ, it is important to distinguish between scientific research based on question-centered "disciplines" and research not so based (see also Burian 1992 and, for some hints on institutionalization, Chapters 2 and 5 in this volume). Disciplines in this sense are obviously and importantly connected with favored, though changing, bodies of technique and practice and are marked by a variety of sociological indicators: journals, departments, communication networks, pedagogical tools, Kuhnian exemplars, patronage, and so on. But, the sociological indicators are not enough: If one does not consider the central questions of a discipline, one's understanding of the dynamics of disciplinary and scientific change will be defective. Again, this point is not terminological. If we fix the use of the term "discipline" as I have suggested, skeptics may still challenge the usefulness of the category *discipline*, thus understood, for describing the organization of scientific work and practices.

*Prima facie*, molecular genetics is a discipline in the sense just adumbrated. It is built around a small body of central questions: What is the genetic material? How is that material organized? How does gene structure relate to gene expression and gene function? By what mechanisms are genes (or is genetic information) transmitted from one generation to the next? How does the genetic material eventually affect particular traits of organisms? For example, given all the right cellular machinery, how do genetic differences (interacting with other contributing causes) bring about sexual differentiation or determine whether a cell is able to respire? Because these questions have been pushed to the molecular level, almost all detailed issues in molecular genetics are finer-grained than these exemplary questions suggest, but most work in that discipline can be easily located with respect to just such questions.

These exemplary questions are not written in stone for they are transformed by changing practices, techniques, beliefs, and knowledge. This is readily seen by comparing the technical instantiations of the central questions of genetics at different times and places. Compare, for example, early-twentieth-century Mendelian formulations regarding transmission of unit characters with those of the 1930s and 1940s regarding autocatalysis and heterocatalysis, or those of the 1960s and 1970s regarding DNA replication and transcription, gene expression, and post-transcriptional processing. Or, again, compare the emphasis in France just after mid-century on genetic regulation of cellular states with the contemporaneous emphasis in the United States on the structure of the genetic material. Still, the core issues that run through all of these alternative formulations were near the center of the discipline of genetics for most of the last century. They concern the materials of which genes (or units of genetic function) are made, their physical structure and organization, how they are transmitted, how they function, and how their functions are controlled.

This chapter focuses particularly on one way in which questions are transformed. Questions are refracted and altered by the interaction between what a number of us have called "local traditions" or "local cultures" (see, e.g., Gaudillière 1992 and 1993). I amplify some of the ways in which these local traditions interact, but first let me complete the contrast between molecular biology and molecular genetics as just characterized.

Not all biological work is organized in question-centered disciplines like genetics. Thus, taxonomic branches of biology – from botany and zoology to algology, ornithology, and protozoology – are not organized around focal questions like those of genetics. Molecular biology – by which I understand the study of organisms and biological processes at the molecular level – is more like molecular botany or molecular zoology than it is like molecular genetics, more like molecular auto mechanics than molecular mechanics. To be sure, molecular biology mainly studies molecular mechanisms (Burian 1996). But, consider the variety of mechanisms at stake: protein folding; protein action and interaction; muscle and nerve action; fluid transport in phloem and xylem; metabolite action; kinetosome (re?)production; the molecular basis of immune reactions; the interactions among receptors, hormones, and hormone analogs; and many aspects of photosynthesis (Zallen 1996), bioenergetics, biochemistry, and so on.

This list is somewhat random; molecular biology deals with an immense diversity of issues. In consequence, there are important questions about the glue that holds this so-called discipline together. My own view is that the best way to understand the interactions across the different problems, disciplines, and traditions that are subsumed under molecular biology is in terms of the overlapping techniques that allow many questions – posed by different disciplines and by investigators in different contexts – to be approached by common tools and with a variety of experimental systems.

These ideas about the nature of molecular biology raise important issues about the ways in which techniques are passed from group to group, from discipline to discipline – and the ways in which they are transformed in the process. The central theme of this chapter is that disciplinary dynamics are greatly affected by the importation of novel experimental systems, techniques, and questions from "outside" (i.e., from other disciplines and groups) into working groups and local traditions. My thesis is that *the technical instantiations of the central questions of a discipline are often transformed by the attempts of working groups to domesticate novel techniques, perspectives, and experimental systems in the attempt to address those central questions, as yet unresolved within the preexisting "local" knowledge context.* This thesis has consequences regarding the proper mix of sociological and "object-level" (experimental or theoretical) analyses needed to understand disciplinary dynamics. That mix will be the focus of a subsequent study; this chapter outlines paired case studies that support the thesis and illustrate its application.

For this purpose, I examine certain aspects of the work on protein synthesis in the laboratories of Jacques Monod and Paul Zamecnik. One aim in doing so is to stimulate others to develop a better account than I can provide of the modes of interchange and influence between experimental and theoretical workers, between those who have command of a technique and those who have a problem to which to apply it, between workers in different disciplines. One of the important tasks for the history of science during the next decade is to clarify the structure, substance, and effects of such interchanges.[2]

[2] In a number of recent lectures, Peter Galison has discussed this issue in terms of a "trading zone" between experimental and theoretical workers. I believe a more general approach and terminology are needed because the interchange takes place between working groups of many stripes and is not restricted to any particular location or "zone." The present study illustrates these points. An example, not discussed in this chapter, may reinforce this point: it is the importation of gel electrophoresis into population genetics from protein biochemistry (Hubby and Lewontin 1966; Lewontin and Hubby 1966). The practice of the former discipline was thoroughly transformed in the process, but it is highly strained to treat the adoption of a thoroughly routinized practice from another context in terms of interactions in a "trading zone."

## DIFFERENTIATION: PROTEIN SYNTHESIS IN PROCARYOTES AS A MODEL

Some of the founding figures of molecular genetics became interested in the problems surrounding genetic control of cell differentiation and organismic development around the time of World War II. During the 1950s and 1960s, many of them treated differentiation in terms of the regulation of protein synthesis. The underlying hypothesis was that differentiation is an irreversible commitment of a cell lineage to the manufacture of a coordinated set of "luxury" proteins – that is, specialized proteins not needed to maintain the life of the cell. Thus, the primary differences among nerve, kidney, skin, and blood cells were thought to depend on the specialized sets of proteins that they make, which, in turn, affect their morphologies, interactions with other cells, and responses to biological signals and stimuli.

With this approach, the problems of control of gene expression and protein synthesis were substituted for the hitherto intractable problem of differentiation. For reasons of technical ease, many founders of molecular genetics came to study protein synthesis and its regulation mainly in microorganisms, especially bacteria (procaryotes), yeasts, and single-celled fungi (eucaryotes) rather than in multicellular eucaryotes. I cannot go into great detail here, but even a minimal sketch of some aspects of this substitution of the difficult but relatively more tractable problem of control of protein synthesis in microorganisms for the problem of differentiation in multicellular eucaryotes proves illuminating.

To capture the spirit of this substitution, I quote from Jacques Monod's first major English-language paper, a review of work on enzymatic adaptation. Enzymatic adaptation concerns the ability of starved bacteria to switch from a depleted carbon source to an entirely different carbon source. It was renamed "enzyme induction" in the mid-1950s, after it had been established that bacteria make a new battery of enzymes to digest a new carbon and that they possess but do not express the genetic information for making those enzymes before switching.[3] In 1947, Monod justified attention to this then-esoteric topic by arguing for its larger biological significance. He wrote: "The widest gap, still to be filled, between two fields of research in biology, is probably the one between genetics and embryology. It is the repeatedly stated – and thus far unsolved – problem of understanding how cells with identical genomes may become differentiated, that of acquiring the property

[3] It is particularly interesting to compare Spiegelman and Monod in this connection, as is done by Gaudillière (1992) and by Gilbert (1996). Cf. also Monod et al. (1953).

of manufacturing molecules with new, or at least, different specific patterns or configurations."[4]

In the 1960s, Monod made the notorious remark that what is true for *E. coli* is true for the elephant. In dealing with this claim, it helps to recall how central the problem of differentiation had been in his thinking. *Monod substituted a technically well-articulated problem for a prior, ill-defined general problem:* Overstating, but not by much, he substituted the problem of enzyme induction (or, more generally, control of protein synthesis) in the bacterium *E. coli* for that of differentiation in the elephant and all multicellular organisms.

In making this substitution, Monod's group was bringing distinct disciplines and local traditions together. The problem of differentiation had belonged to embryology and physiological genetics, that of enzymatic adaptation to bacterial physiology and biochemistry. Furthermore, because enzymatic adaptation does not result in overt differentiation but only in a controlled shift in the enzymes manufactured by bacterial cells, the substitution of the problem of enzymatic adaptation, then enzyme induction, then protein

[4] (Monod 1947, p. 224). The persistence of this theme in Monod's work is obvious from the conclusion of his and Jacob's review of work on the operon and protein synthesis (Jacob and Monod 1961).

> The conclusions apply strictly to the bacterial systems from which they were derived; but the fact that adaptive enzyme systems of both types (inducible and repressible) and phage systems appear to obey the same fundamental mechanisms of control, involving the same essential elements, argues strongly for the generality of what may be called "repressive genetic regulation" of protein synthesis . . .
>
> The occurrence of inductive and repressive effects in tissues of higher organisms has been observed in many instances, although it has not proved possible so far to analyze any of these systems in detail . . . It has repeatedly been pointed out that enzymatic adaptation, as studied in micro-organisms, offers a valuable model for the interpretation of biochemical co-ordination within tissues and between organs in higher organisms. The demonstration that adaptive effects in micro-organisms are primarily negative (repressive), that they are controlled by functionally specialized genes and operate at the genetic level, would seem greatly to widen the possibilities of interpretation. The fundamental problem of chemical physiology and of embryology is to understand why tissue cells do not all express, all the time, all the potentialities inherent in their genome. The survival of the organism requires that many, and, in some tissues most, of these potentialities must be unexpressed, that is to say *repressed*. Malignancy is adequately described as a breakdown of one or several growth controlling systems, and the genetic origin of this breakdown can hardly be doubted.
>
> According to the strictly structural concept, the genome is considered as a mosaic of independent molecular blueprints for the building of individual cellular constituents. In the execution of these plans, however, co-ordination is evidently of absolute survival value. The discovery of regulator and operator genes, and of repressive regulation of the activity of structural genes, reveals that the genome contains not only a series of blue-prints, but a co-ordinated program of protein synthesis and the means of controlling its execution (p. 354).

synthesis, for that of differentiation in the cells of higher organisms was quite adventuresome.

Such substitutions often stem from interactions between local traditions in which competing groups articulate their questions and develop their techniques so as to make their work more effective than their competitors'. The immediate traditions within which Monod began his work were those of microbiology in the Pasteur Institute, as instantiated in André Lwoff's nutritional biochemistry, and of Boris Ephrussi's program to reconcile embryology with genetics.[5] The interactions with these traditions (which were by no means the only ones that Monod exploited) transformed both of them in important ways and were resisted by both Ephrussi and Lwoff. Ephrussi resisted the assimilation of differentiation to gene regulation in procaryotic cells because he was convinced that cytoplasmic heredity in different cell lineages caused differentiation in eucaryotes[6] whereas, well into the 1950s, Lwoff maintained that *cellular* regulation involved elements "endowed with genetic continuity" located throughout the cell, and that it was quite distinct from genetic regulation, especially the regulation of protein synthesis (Lwoff 1949; see also Burian and Gayon 1991).

Monod and his coworkers employed the techniques of chemical kinetics; these techniques set their work off from that of those who worked mainly with cell structure, end products, or biochemical equilibria. The combination of his commitments regarding technique and his views about differentiation helps explain important aspects of his experimental program. Although he learned to do genetic experiments in the late 1930s at Caltech and although he often argued from the mid-1940s on that genetics is crucial to understanding the control of protein synthesis, he did almost no genetic experiments

[5] These claims are controversial. Jean Gayon, Doris Zallen, and I have examined these traditions and provided some evidence of Monod's allegiance to them (Burian 1990; Burian and Gayon, 1991; Burian, Gayon, and Zallen 1988, 1991). Gaudillière's dissertation (1991), which pays much greater attention than we have to the institutional and biochemical settings relevant to Monod's work, reinforced our view on this issue. Gaudillière now rightly emphasizes the importance of a third tradition, that of Pasteurian immunochemistry, in Monod's work (Gaudillière 1993).

[6] Ephrussi argued, cautiously, that one could not extrapolate safely from procaryotes to eucaryotes, but he did not insist that the mechanisms he favored would be found. Well before he started working in somatic cell genetics, however, he insisted on the need to study differentiation in multicellular eucaryotes. For arguments that differentiation may be caused in the cytoplasm, see, for example, Ephrussi 1953, especially pp. 4–6 and 99–109). Later, in Ephrussi 1956, Ephrussi combated the orthodoxy of the day by insisting that differentiation should be treated nearly independently of microbiological findings because the most striking phenomena of differentiation occur in metazoa – and that (following Lederberg), "embryology will ultimately have to be studied in embryos." Cf. also Ephrussi (1958), Lederberg (1958, especially pp. 384–8), and Lederberg and Lederberg (1956, especially pp. 113–15). For an account of Ephrussi's views on this topic as he turned to work on somatic cell genetics, see Zallen and Burian (1992).

until 1958,[7] finally employing techniques brought into his group by Pardee and Jacob (Pardee, Jacob, and Monod 1958, 1959). The reason is straightforward: Classical genetics provided no approach to fundamental questions about differentiation and development and no easy way to do kinetic experiments. When he adopted (and adapted) the genetic tool, it was in kinetic experiments – possible only in microorganisms – utilized to analyze the timing of gene expression via the production of proteins under the control of a newly introduced gene. It is only at this point that Monod's personal research can be said to have begun dealing directly with traditional genetic questions.

It is too large a job to explore here the extent to which molecular geneticists substituted the study of protein synthesis and its control for the study of differentiation during the 1950s and 1960s. However, an impressionistic survey of a number of key papers during this period shows that an important group of leaders in the molecularization of genetics thought that differentiation could be fully explained in terms of gene expression as analyzed in studies of protein synthesis.[8] After all, most of them were committed to the view that all information relevant to development is encoded in chromosomal DNA (perhaps allowing minor exceptions for organelles like chloroplasts and, later, mitochondria). There were no good tools with which to look at post-transcriptional processing of mRNAs until well into the 1970s – and, even if there had been, the predominant ethos among geneticists (as opposed to biochemists or embryologists) would have treated post-transcriptional processing as part of the genetically controlled apparatus for regulating gene expression.

Given the tools available in the 1950s, microorganisms offered great technical advantages for the study of protein synthesis and its genetic control. It is not surprising that the first steps toward many of the breakthroughs alluded to previously were made by the use of microorganisms. Among the favored organisms, procaryotes like *E. coli* K-12 and single-celled eucaryotes like *Neurospora* and yeast each offered specific advantages after extensive domestication. This is not the place to discuss the advantages and liabilities of particular experimental systems, but each of them contributed significantly

[7] There is room to argue about which experiments to count as genetic. An important borderline case, perhaps the only exception, is Monod (1950), which shows that genes and enzymes are distinct entities because the presence of a gene allowing facultative production of an enzyme is independent of the actual *production* of the enzyme. But, the focus of Monod's concern even here is not centrally genetic: it is the role of the substrate in determining the "specific structure" of adaptive enzymes "in the production the specific molecule" (p. 58).

[8] A single emblematic reference illustrates this theme: Markert (1964).

and distinctively to the reformulation of questions about the control of protein synthesis and to the development of new experimental tools.[9]

## GENETICS, BIOCHEMISTRY, AND PROTEIN SYNTHESIS: *E. COLI* VERSUS TISSUE SLICES AND CELL-FREE SYSTEMS[10]

Monod was far from alone in substituting the analysis of protein synthesis plus control of gene expression for an analysis of differentiation. Yet, it is important to recognize that some workers took on protein synthesis as such, whereas others (like Monod) took it on both in its own right and as part of a further set of problems involving gene expression and differentiation. I now illustrate some differences in problem articulation, technique, and choice of experimental system that arose among those concerned with gene expression and some of the biochemists working on protein synthesis. This discussion illustrates, the roles of local tradition, technique, and experimental system in shaping problems and altering questions.

In the 1950s, biochemists studying protein synthesis in liver tissue slices developed a cell-free system in which to study protein synthesis. The very fact that creating a cell-free system was a primary objective reveals a central difference with Monod: Such a system served Monod as a useful incidental tool, but it could not solve his main problems concerning cellular mechanism, regulation, and control of protein synthesis.

The biochemical work just alluded to made fundamental contributions to our understanding of protein synthesis. We consider here important work carried out in the mid-1950s in Paul Zamecnik's laboratory, one of the John Collins Warren laboratories of the Collis P. Huntington Memorial Hospital of Harvard University.

The Huntington laboratories were directed by Joseph Aub from 1936 to 1956, in which role Aub was succeeded in 1956 by Zamecnik. Aub, who had a formative influence on Zamecnik's career, believed that cancer is a form of unregulated or improperly regulated growth; thus, he established a local tradition of basic research on cancer in the context of studies of (normal)

[9] A useful account of the particular advantages offered by *Neurospora* is presented by Perkins (1992). It is to be hoped that more studies (especially comparative ones) of the advantages – and consequences – of domesticating particular organisms will be carried out in the near future.

[10] See Rheinberger (1992a, 1992b, 1993, 1995) for close studies of the Zamecnik group's work on ribosomes. See also Hoagland (1990, especially Chapters 4–6) for a useful retrospective account of the same work.

growth. Accordingly, he encouraged Zamecnik to pursue work on the mechanisms of protein synthesis independent of any particular focus on cancer as such.[11] In the work at issue here, Zamecnik's group established the existence and importance of RNAs initially classified operationally as "soluble" RNA (sRNA), now called transfer RNA (tRNA): sRNA was the RNA still in solution in the supernatant at pH 5 after "microsomes" – ribosomes, fragments of endoplasmic reticulum, and related material – were spun out at 100,000 g in an ultracentrifuge. A cell-free system had to include sRNA to make new protein. Zamecnik's group showed by biochemical techniques that this RNA was involved in the "activation" of amino acids (a step previously identified as necessary before amino acids could be incorporated into proteins) and in transport of the activated amino acids to the ribosomes, where they were incorporated into proteins.

By 1958, Zamecnik's laboratory had demonstrated that there are many different soluble RNAs, all relatively small (i.e., about 70 or 80 nucleotides) and that some – probably all – of them combine enzymatically with a single activated amino acid. In brief, these RNAs were the "adaptors" of Crick's "adaptor hypothesis,"[12] the vehicles by which single amino acids were brought to the ribosome to be added to the growing string of amino acids synthesized there.[13] They soon knew, furthermore, that the sRNAs, each small and bound to a single amino acid, could not determine the sequence of the amino acids synthesized by the cell.

Work with the cell-free system and many others demonstrated that the RNA in ribosomes is tightly bound to protein and turns over slowly. By about 1960, in light of various kinetic analyses plus work on Crick's sequence hypothesis, it became unlikely that the signal containing the information for building a new protein is in the RNA of the ribosomes.[14] Some other signal

[11] For background, see Hoagland (1990) and Rheinberger (1992a, 1992b, 1993).

[12] See Judson (1979, pp. 268–70, 287–95, 313–28, et passim). See also Hoagland (1960) and the portion of Crick's originally unpublished "On Degenerate Templates and the Adaptor Hypothesis," printed at pp. 290–1 in Judson (1979). For retrospective accounts, see Crick (1988) and Hoagland (1990).

[13] Hoagland, Zamecnik, and Stephenson (1959) and Zamecnik (1960, 1969). It is worth adding that Crick thought that the adaptors would have to be considerably smaller than 70–80 bases long; much work was needed to establish that biochemically identified sRNAs served as Crick's postulated adaptors.

[14] A great many pathways led to this result. This stands out in Chantrenne (1961, especially pp. 63 ff.). Here are two examples: McQuillen, Roberts, and Britten (1959), which shows that the turnover of RNA in ribosomes is much slower than that of the signal controlling the production of specific proteins; and Brenner (1961), which shows that phage use preexisting bacterial ribosomes to manufacture phage protein. The experimental arguments to this effect depended less on the specifics of any single set of experiments than on the accumulation of a great variety

had to travel from the DNA to the ribosomes for the DNA to determine the order of the amino acids in the proteins. This signal, found and analyzed in considerable detail in the first half of the 1960s, came to be called messenger RNA (mRNA). It was found by a kind of triangulation involving the work of many groups working with different tools and in different traditions.[15]

One reason for putting the matter this way is to counteract the heroic version – common to Crick, Horace Judson, and many others – according to which the key insight belonged to a small group of insiders from Cambridge, the Institut Pasteur, and the phage group.[16] On the contrary, it was the interactions within a much wider network that made the existence of mRNA seem plausible, even necessary. Let me reinforce this point.

There is considerable pre-1960 evidence, supported by Zamecnik's spontaneous recollections in 1979 and Mahlon Hoagland's retrospective account,[17] to show that the early work on sRNA was wholly independent of Crick's theorizing. Until 1957, the Zamecnik group's development of the liver slice and cell-free systems had no connection at all with Crick's adaptor or sequence hypotheses, which were first stated publicly after those systems had been applied to the problems of amino-acid activation and transport to the ribosome. This independence of investigative pathways is in keeping with Watson and Crick's early avoidance of biochemistry and the firmly biochemical orientation of Zamecnik's work.[18] Indeed, it is a familiar point – one that I will

of information about gene action, rapidity of signal generation, speed of processing, speed of turnover, interchangeability of tRNAs across systems, similarities among ribosomes, and so on. For this reason, though the PaJaMo experiment (Pardee, Jacob, and Monod 1959) was crucial in calling forth the hypothesis of mRNA, it was but one of many pathways toward recognition of the need for a new class of RNAs.

[15] There are literally thousands of relevant references, including Gros et al. (1961); Hershey (1953); Hoagland et al. (1958); Lamborg and Zamecnik (1960); Nirenberg and Matthei (1961); Nomura, Hall, and Spiegelman (1960); Rich (1960); and Volkin and Astrachan (1956). Cf. Watson's Nobel Lecture Watson (1963) for a good contemporary review and Siekevitz and Zamecnik (1981) for a biochemical retrospective, oriented mainly toward work on ribosomes and the role of tRNAs.

[16] See, for example, Judson (1979, pp. 427 ff.). Many variants of this account have been published: for example, Crick (1988, Chapter 11).

[17] See Zamecnik's response to Olby in Zamecnik (1979, pp. 299–300). On the independence of the work in Zamecnik's lab from that of Crick's adaptor hypothesis as of 1957, see Hoagland (1990, p. 93 ff.); Hoagland, Zamecnik, and Stephenson (1959, especially pp. 112–13); and Zamecnik (1969). This impression is strengthened by reference to the bibliographies of Zamecnik's pre-1960 papers. Judson (1979, pp. 326–7) provides more evidence that Zamecnik and Hoagland had completed their 1956 experiments yielding attachment of activated hot leucine to sRNA, followed by its incorporation into a protein, before they learned of the adaptor hypothesis.

[18] Thus, Zamecnik (1960) treats protein synthesis in terms of the biochemistry of "the pathway from free amino acid to biologically active protein" (p. 256). His concern is mechanisms of synthesis: The analysis of protein synthesis might yield understanding of the mechanisms of peptide bond synthesis; the reversed action of proteolytic enzymes cannot explain the precisely

soon reinforce – that the breaking of the genetic code was, in the end, largely a *biochemical* achievement, accomplished primarily by biochemical and not molecular–biological techniques. Controversially, the same can be said for the run up to the discovery of mRNA.

A helpful way of understanding the contrast implied here between biochemical and molecular–biological traditions concerns the way the cell was analyzed into functional parts. For Zamecnik and the biochemists in his group, cells were analyzed into compartments in terms of the *operations* that separated bits of cellular machinery. Thus, there was the mitochondrial compartment (defined in terms of the pellet that included mitochondria under low-speed centrifugation), ATP and the ATP production system (the biochemically isolated energy-production system), the microsomal compartment (which included the materials sedimented by high-speed centrifugation), and soluble RNA (plus whatever other substances were left after the purification procedures applied to the final supernatant following high-speed centrifugation).

In contrast, Monod's group, which was concerned with the regulation of cellular states and the physiological controls that determine which enzymes are synthesized by a cell, divided cells in terms of cellular geography (e.g., cytoplasm versus chromosome) or functional geography, as measured by *in vivo* kinetics. Their compartmentalization did not match the one produced by Zamecnik's operations. They did not study the action of inducing agents on isolated ribosomes, and their use of cell-free extracts served largely to support the analyses produced by *in vivo* kinetics. Applied to enzyme induction, that tool was supposed to reveal where inducing agents acted and which mechanisms were activated to produce the substances whose synthesis was induced.

This contrast between Monod and Zamecnik's groups marks a series of technical commitments and differences in their central problems. The Zamecnik group, seeking to understand normal and abnormal growth in the tradition of Aub, was most concerned with the machinery by which the cell manufactures proteins. The Monod group was most concerned with the regulatory system by means of which cells managed to alter or switch the enzymes or proteins they manufactured, with the primary aim of solving the problems surrounding differentiation rather than those surrounding growth.

defined sequence of amino-acid incorporation into protein. Soluble RNA supplies activated amino acids to the machinery that assembles proteins; rRNA is a constituent of that machinery. RNA also has something to do with condensing the amino acids in the correct order in a synthesized protein, but there is no evidence that Zamecnik's attack on this problem prior to 1960 depended significantly on recognition of the informational or genetic aspects of nucleic acids. He viewed the problem in terms of the biochemistry of nucleic acids and proteins.

If both groups could be said to be working on "the" problem of protein synthesis, that problem was importantly transformed by the local traditions through which it was viewed.

## THEORY, EXPERIMENT, AND EXCHANGE ACROSS BOUNDARIES: SOME COMMENTS ON CRACKING THE GENETIC CODE

The dead ends that Crick encountered in his beautiful theoretical papers attempting to crack the genetic code are of particular interest for our purposes. Crick posed the variants of the coding problem so clearly that it is slightly less difficult than it might otherwise be in so complex a case to trace the tangled pathways by which the issues he raised were resolved. As is well known, those pathways were quite different from the ones that he first thought would bear fruit: They were far more biochemical and far less theory-driven than he anticipated.[19] A cursory examination of the differences between Crick's early expectations and the experimental pathways by means of which those questions were answered is revealing.

The main point to be gained from such considerations does not require much detail. It is that the mRNAs that determine the sequence of amino acids in proteins and the correlating code were far more likely to be discovered by means of a biochemical detour than by means of structural analyses like those attempted by Watson as extensions of Watson and Crick's analysis of DNA (e.g., Tissiéres and Watson 1958, reviewed in Watson 1963) or by means of theoretical analyses of coding constraints like those attempted by Crick. Put strongly: Because of the nature of the coding apparatus – *which was, of course, unknown* – structural and in-principle theoretical solutions to the coding problem were beyond the available means – and remain so even now.

The fact that the code is in a strong sense arbitrary could only be found out experimentally. The fact that it is not based, for example, on a spacing mechanism such that each amino acid fits into a unique pocket of information-bearing DNA or RNA means that Watson's structural approach had no point of

[19] The standard reference here is Nirenberg and Matthei (1961). See also Judson (1979). By 1963, Crick's judgment on "the place of theory" was rather tempered: Good theories, even if incorrect, "enable us to tighten up our logic and make us scrutinize the experimental evidence to some purpose . . . Whether theory can help by suggesting the general structure of the code remains to be seen. If the code does have a logical structure, there is little doubt that its discovery would greatly help the experimental work. Failing that, the main use of theory may be to suggest novel forms of evidence and to sharpen critical judgment. In the final analysis it is the quality of the experimental work that will be decisive" (Crick 1963, pp. 213–14).

purchase. Such theoretical approaches as those of Crick and George Gamow (1954; see also Judson 1979, pp. 255 ff.) are unworkable because of the mechanics underlying the arbitrariness of the code. As Zamecnik's (and others') work showed by 1960, there are two sites on tRNAs that can vary independently of each other: one for the encoding sequence (soon shown to be, as Crick suggested, a triplet) and one for the attachment of a single amino acid.[20]

Furthermore, there is no constraint barring redundancy in the code. Taken together, these results have a consequence that could not have been known *a priori: No abstract treatment of the coding problem can produce a determinate solution.*

Furthermore, because proteins are not directly synthesized on a DNA template, to get at the code it is necessary to operate on the intermediate that carries sequence information. Because this intermediate, mRNA, is usually short-lived, unstable, and a relatively minor proportion of cellular RNA, in practice one had to develop mutually adapted experimental systems and biochemical protocols to do that job. The requirements for getting at the mechanisms of protein synthesis and information transfer by this route could not have been foreseen.

This account suggests that, given the available techniques, the unanticipated biochemical detour was unavoidable in practice. There is a general lesson here for students of creative science: Most interesting problems in science are open ones, subject to reformulation. It is advisable (perhaps necessary) for work on such problems to take place via exchange across the boundaries of disciplines, groups, and local traditions. Opportunism is the rule in tackling open problems because they are likely to require mobilization across interfaces – mobilization of techniques, traditions, disciplines, and – in the case of experimental biology – a variety of experimental systems.

Attempts to solve open problems must be disciplined by creating reliable practices potentially bearing on those problems. The development of such practices is typically the accomplishment of small groups of laboratory specialists working with some degree of institutional protection within a discipline or a local tradition. But, reliable practices are not enough; reliable practices, after all, may reliably produce artifacts. It is also necessary to show by triangulation that those practices answer the original problem or a well-established successor. *Creative science requires not only severe technical*

[20] The unanticipated independence of these sites provides one reason that Crick underestimated the size of the adaptors. Evidence clinching their independence was provided by Chapeville et al. (1962), which showed that replacing the amino acid on a particular tRNA by a different amino acid substitutes the new amino acid into proteins at the precise places where the original amino acid occurs when the unaltered tRNA is used.

*refinement of specialized tools and techniques within local traditions but also leaky boundaries allowing exchange between groups and traditions, so that the competing problems, techniques, and questions can interpenetrate each other, so that triangulation is possible.*

## MORE ABOUT CHOICE OF TECHNIQUE AND EXPERIMENTAL SYSTEM

In consequence of these considerations, the matching of experimental system to techniques is critical. In general, one cannot know in advance what steps will yield a good match. For example, as could not then have been known, in the late 1950s a tissue culture of heart or muscle cells would not have yielded enough tRNAs or mRNAs to be detected by the techniques then in use, especially given the uncertainties about the details of protein synthesis. Again, the liver-slice system and the cell-free-culture system based on it contained so many proteases (i.e., enzymes that digest proteins) that, even though it was possible to chase activated amino acids attached to tRNAs as far as the ribosomes, it was extremely difficult to follow the incorporation of amino acids into synthesized protein.[21] Because the work was at the edge of available technique *and because it was unclear what was being sought*, many artifacts slowed the analysis of the details of protein synthesis and the recognition of messenger RNA.[22] Additionally, the kinetics of many experimental systems were not clean enough to make the clues they offered stand out. This was the case for Elliot Volkin and Lazarus Astrachan's (1956) new RNA fraction in *E. coli* after infection with phage T2. Their RNA had a base-pair composition matching the phage's DNA, and it turned over quickly. In hindsight, they had

[21] The creation of an *E. coli*–derived cell-free system – sought earlier, but finally achieved in 1960 (Lamborg and Zamecnik 1960) – provided a crucial tool used by Tissières, Watson, Matthei, Nirenberg, and many others during the period of most rapid progress in the early 1960s. Further work is needed to assess the advantages of this system over its predecessors, but it clearly proved to be of major importance.

[22] It is unusual for scientists to acknowledge their near misses openly. Zamecnik offers a good example of such an admission: "*Description of a blind spot*. Let me mention one of our efforts to examine the polypeptide polymerization site by means of electron microscopy . . . In 1960 we began to prepare ribosomes from rabbit reticulocytes which [we] subjected to high resolution microscopy. We looked hard for our principal objective, the location of tRNA molecules on the ribosomes, but without success. Frequently there were to be seen strands of what appeared to be RNA unraveling from the ribosomes and running from one ribosome to another . . . I felt uneasy about this, considering it to be evidence of roughness in our preparative methods. A year or two later these electron micrographs might have served as textbook pictures of messenger RNA and polysomes. However, at the moment I was fixed on the importance of ribosomal RNA itself as the genetic message [as was nearly everyone in the field! – RB] and had no eyes for the messenger" (Zamecnik 1969, p. 9).

found a fairly clean mRNA signal but, at the time, it was not clear that it was connected with protein synthesis. Volkin and Astrachan themselves suggested that they had found an intermediate in the manufacture of phage DNA, not of protein.

In the later 1950s, in virtue of the then-unknown details of protein synthesis, biochemical kinetics turned out to be one of the most useful of the available techniques for studying the intermediate steps between DNA and protein. Biochemical kinetics were not part of the armamentarium with which Crick or most members of the phage group initially approached the problem. There was nothing wrong with their choice to place their bets elsewhere – provided that, like Crick, they paid attention to developments that forced the problem onto new turf and, in the process, transformed it. What makes this result all the more revealing is that the new turf was already occupied by preexisting groups that had not yet recognized the aptness of their tools for a problem that they had not considered – especially not in Crick's style.

In circumstances like these, the interactions among groups with flourishing local traditions – groups flexible enough to allow their tools and problem formulations to be bent by those interactions – are crucial to the dynamics of disciplines and of scientific knowledge and to the formation of new disciplines such as molecular genetics. When potentially influential groups working at interdisciplinary interfaces revise prior practices and problems, they create the opportunity to establish new traditions, perhaps even new disciplines. To use an analogy familiar from evolutionary biology, it is in such circumstances that founder effects occur. These play a crucial role in the founding of major new traditions.

## TASK DEFINITION

Choice of technique and experimental system are not enough. There is also the problem of task definition. The direction in which a line of work is taken is sensitive to the questions with which one starts and to the connections of those questions with the central problems of the disciplines to which the relevant investigators are (or come to be) allied. There are powerful effects here, as can be seen by the contrast between Monod's and Zamecnik's groups, both of which employed chemical kinetics on similar systems. Someone centrally interested, like Monod, in *the regulation of protein synthesis* will make different choices than someone investigating, like Zamecnik, *the mechanics of the assembly of the chain of amino acids*. These differences greatly affected the "investigative pathways," to use Holmes's phrase, of the Monod and Zamecnik groups.

Thus, although the choices of the groups overlapped, one's signal was, at least sometimes, the other's noise. The groups continued to employ different systems – the former aiming to understand the genetic and epigenetic *control* of protein synthesis, the latter to isolate and characterize the *machinery* that assembles proteins and the signals it employs. Their choices of experimental systems and of focal issues had major effects, as such choices often do, driving the groups in different directions. Thus, the "natural" job to do with an intact cell that manufactures different enzymes in precisely determined circumstances is to work out the ways in which those commitments are made or controlled, whereas the "natural" job to do with a cell-free system is to work out the details of the machinery assembling proteins. The differences in the jobs thus led to different understandings of common issues and to the elaboration of different tools, concepts, and issues (e.g., allostery versus ribosome biochemistry, structure, and function). Such differences are the stuff of which local traditions are made – and they make it necessary for creative scientists to reach out beyond their local traditions.

## CONCLUSION

Our central case studies suggest that if one wishes to understand the approaches that laboratory groups take toward major problems such as protein synthesis or the regulation of gene expression, it is important to articulate the issues of concern in the local cultures within which those groups operate. They also suggest that to get at the dynamics of disciplinary change – indeed, of scientific change generally – it is equally necessary to understand the interactions *between* laboratory groups and between local scientific cultures. The ways in which major scientific problems are handled and evolve are, of course, affected by a great many factors. I have shown that these include at least the uses that laboratory groups make of each other's findings, the power and limitations of the techniques they employ, the idiosyncrasies of the various experimental systems they employ, and the transformations that all of these effect in the technical articulation of the central questions of the relevant discipline(s). I hope that this chapter encourages others to take up the challenge of articulating the interactions among all these factors.

In closing, I would like to raise a larger historiographical issue that indicates how difficult this job may prove to be. The issue is the problem of the scale on which we work. One can examine individual careers, laboratory groups, local traditions, communication networks, institutions, disciplines, national traditions and cultures, and problems that are continued for long periods across boundaries of all these sorts. It is obviously important to engage in comparative studies at all of these levels because each level affects the dynamics of scientific institutions, disciplines, theories, and local practices. Yet, there remains a nagging worry about how to integrate the findings of such diverse studies because one of the most difficult problems facing students of scientific work and scientific processes is that of evaluating the interaction of factors across such levels. It seems to me that one of the most important tasks for historians working with case studies like those I have sketched herein is to confront us with that problem and to force us to face it squarely.

## REFERENCES

Brenner, S. 1961. "RNA, ribosomes, and protein synthesis." *Cold Spring Harbor Symposia on Quantitative Biology* 26: 101–10.

Burian, R. M. 1990. "La contribution française aux instruments de recherche dans le domaine de la génétique moléculaire." In *Histoire de la génétique: Pratiques, techniques et théories*, eds. J.-L. Fischer and W. H. Schneider. Paris: ARPEM, 247–69.

Burian, R. M. 1992. "How the choice of experimental organism matters: Biological practices and discipline boundaries." *Synthese* 92: 151–66.

Burian, R. M. 1996. "Underappreciated pathways toward molecular genetics as illustrated by Jean Brachet's chemical embryology." In *The Philosophy and History of Molecular Biology: New Perspectives*, ed. S. Sarkar. Dordrecht, Holland: Kluwer, 67–85.

Burian, R. M., and J. Gayon. 1991. "Un évolutionniste Bernardien à l'Institut Pasteur? Morphologie des Ciliés et évolution physiologique dans l'oeuvres d'André Lwoff." In *L'Institut Pasteur: Contribution a son histoire*, ed. M. Morange. Paris: Editions la Découverte, 165–86.

Burian, R. M., J. Gayon, and D. T. Zallen. 1988. "The singular fate of genetics in the history of French biology, 1900–1940." *Journal of the History of Biology* 21: 357–402.

Burian, R. M., J. Gayon, and D. T. Zallen. 1991. "Boris Ephrussi and the synthesis of genetics and embryology." In *A Conceptual History of Embryology*, ed. S. Gilbert. New York: Plenum, 207–27.

Chantrenne, H. 1961. *The Biosynthesis of Proteins*. Oxford: Pergamon Press.

Chapeville, F., F. Lipmann, G. V. Ehrenstein, B. Weisblum, W. J. Ray, and S. Benzer. 1962. "On the role of soluble RNA in coding for amino acids." *Proceedings of the National Academy of Sciences, USA* 48: 1086–92.

Crick, F. H. C. 1958. "On protein synthesis." *The Biological Replication of Macromolecules: Symposia of the Society for Experimental Biology* 12: 138–63.

Crick, F. H. C. 1963. "The recent excitement in the coding problem." *Progress in Nucleic Acid Research* 1:163–217.

Crick, F. H. C. 1988. *What Mad Pursuit: A Personal View of Scientific Discovery*. New York: Basic Books.

Ephrussi, B. 1953. *Nucleo-Cytoplasmic Relations in Microorganisms: Their Bearing on Cell Heredity and Differentiation*. Oxford: Oxford University Press.

Ephrussi, B. 1956. "Enzymes in cellular differentiation." In *Enzymes: Units of Biological Structure*, ed. O. Gaebler. New York: Academic Press, 29–40.
Ephrussi, B. 1958. "The cytoplasm and somatic cell variation." *Journal of Cellular and Comparative Physiology* 52, suppl. 1: 35–53.
Gamow, G. 1954. "Possible relation between deoxyribonucleic acid and protein structure." *Nature* 173: 18.
Gaudillière, J.-P. 1991. "Biologie moléculaire et biologistes dans les années soixante: La naissance d'une discipline. Le cas français." Université Paris VII.
Gaudillière, J.-P. 1992. "J. Monod, S. Spiegelman et l'adaptation enzymatique: Programmes de recherche, cultures locales et traditions disciplinaires." *History and Philosophy of the Life Sciences* vol. 14, no. 1: 23–71.
Gaudillière, J.-P. 1993. "Molecular biology in the French tradition?" *Journal of the History of Biology* 26: 473–98.
Gilbert, S. F. 1996. "Enzymatic adaptation and the entrance of molecular biology into embryology." In *Philosophy and History of Molecular Biology: New Perspectives*, ed. S. Sarkar. Dordrecht, Holland: Kluwer, 101–23.
Gros, F., H. Hiatt, W. Gilbert, C. G. Kurland, R. W. Risebrough, and J. D. Watson. 1961. "Unstable ribonucleic acid revealed by pulse labeling of *Escherichia coli*." *Nature* 190: 581–5.
Hershey, A. D. 1953. "Nucleic acid economy in bacteria infected with bacteriophage T2. II. Phage precursor nucleic acid." *Journal of General Physiology* 37: 1–23.
Hoagland, M. B. 1960. "The relationship of nucleic acid and protein synthesis as revealed by studies in cell-free systems." In *The Nucleic Acids*, eds. E. Chargaff and J. N. Davidson. New York: Academic Press, 349–409.
Hoagland, M. B. 1990. *Toward the Habit of Truth*. New York: Norton.
Hoagland, M. B., P. C. Zamecnik, and M. L. Stephenson. 1959. "A hypothesis concerning the roles of particulate and soluble ribonucleic acids in protein synthesis." In *A Symposium on Molecular Biology*, ed. R. E. Zirkle. Chicago: University of Chicago Press, 105–14.
Hoagland, M. B., M. L. Stephenson, J. F. Scott, L. I. Hecht, and P. C. Zamecnik. 1958. "A soluble ribonucleic acid intermediate in protein synthesis." *Journal of Biological Chemistry* 231: 241–57.
Hubby, J. L., and R. C. Lewontin. 1966. "A molecular approach to the study of genic heterozygosity in natural populations. I. The number of alleles at different loci in *Drosophila pseudoobscura*." *Genetics* 54: 577–94.
Jacob, F., and J. L. Monod. 1961. "Genetic regulatory mechanisms in the synthesis of proteins." *Journal of Molecular Biology* 3: 318–56.
Judson, H. F. 1979. *The Eighth Day of Creation: Makers of the Revolution in Biology*. New York: Simon and Schuster.
Lamborg, M. R., and P. C. Zamecnik. 1960. "Amino acid incorporation into protein by extracts of *E. coli*." *Biochimica et Biophysica Acta* 42: 206–11.
Lederberg, J., 1958. "Genetic approaches to somatic cell variation: Summary comment." *Journal of Cellular and Comparative Physiology* 52, suppl. 1: 383–401.
Lederberg, J., and E. M. Lederberg. 1956. "Infection and heredity." *Cellular Mechanisms in Differentiation and Growth*.
Lewontin, R. C., and J. L. Hubby. 1966. "A molecular approach to the study of genic heterozygosity in natural populations. II. Amount of variation and degree of

heterozygosity in natural populations of *Drosophila pseudoobscura*." *Genetics* 54: 595–609.

Lwoff, A. 1949. "Les organites doués de continuité génétique chez les Protistes." In *Unités biologiques douées de continuité génétique*. Paris: CNRS, 7–23.

Markert, C. L. 1964. "Cellular differentiation – an expression of differential gene function." In *Second International Congress on Congenital Malformations*, ed. M. Fishbein. New York: International Medical Congress, Ltd., 163–74.

McQuillen, K., R. B. Roberts, and R. J. Britten. 1959. "Synthesis of nascent proteins by ribosomes in *Escherichia coli*." *Proceedings of the National Academy of Sciences, USA* 45: 1437–47.

Monod, J. L. 1947. "The phenomenon of enzymatic adaptation and its bearing on problems of genetics and cellular differentiation." *Growth Symposium* 11: 223–89.

Monod, J. L. 1950. "Adaptation, mutation, and segregation in the formation of bacterial enzymes." *Biochemical Society Symposia* 4: 51–8.

Monod, J. L., M. Cohn, M. R. Pollock, S. Spiegelman, and R. Y. Stanier. 1953. "Terminology of enzyme formation." *Nature* 172: 1096.

Nirenberg, M. W., and J. H. Matthei. 1961. "The dependence of cell-free protein synthesis in *E. coli* upon naturally occurring or synthetic polyribonucleotides." *Proceedings of the National Academy of Sciences, USA* 47: 1588–602.

Nomura, M., B. D. Hall, and S. Spiegelman. 1960. "Characterization of RNA synthesized in *Escherichia coli* after bacteriophage T2 infection." *Journal of Molecular Biology* 2: 306–26.

Pardee, A. B., F. Jacob, and J. L. Monod. 1958. "Sur l'expression et le rôle des allèles 'inductible' et 'constitutif' dans la synthèse de la ß-galactosidase chez des zygotes d'*Escherichia coli*." *Comptes rendus de l'Académie des Sciences, Paris* 246: 3125–8.

Pardee, A. B., F. Jacob, and J. L. Monod. 1959. "The genetic control and cytoplasmic expression of 'inducibility' in the synthesis of ß-galactosidase by *Escherichia coli*." *Journal of Molecular Biology* 1: 165–76.

Perkins, D. D. 1992. "*Neurospora*: The organism behind the molecular revolution." *Genetics* 130: 687–701.

Rheinberger, H.-J. 1992a. "Experiment, difference, and writing. I. Tracing protein synthesis." *Studies in the History and Philosophy of Science* 23: 305–32.

Rheinberger, H.-J. 1992b. "Experiment, difference, and writing: II. The laboratory production of transfer RNA." *Studies in the History and Philosophy of Science* 23: 389–422.

Rheinberger, H.-J. 1993. "Experiment and orientation: Early systems of *in vitro* protein synthesis." *Journal of the History of Biology* 26: 443–71.

Rheinberger, H.-J. 1995. "From microsomes to ribosomes: 'Strategies' of 'representation'." *Journal of the History of Biology* 28, no. 1: 49–89.

Rich, A. 1960. "A hybrid helix containing both deoxyribose and ribose polynucleotides and its relation to the transfer of information between the nucleic acids." *Proceedings of the National Academy of Sciences, USA* 46: 1044–53.

Siekevitz, P., and P. C. Zamecnik. 1981. "Ribosomes and protein synthesis." *Journal of Cell Biology* 91, 3 pt. 2: 53s–65s.

Tissiéres, A., and J. D. Watson. 1958. "Ribonucleoprotein particles from *Escherichia coli*." *Nature* 182: 778–80.

Volkin, E., and L. Astrachan. 1956. "Phosphorus incorporation in *Escherichia coli* ribonucleic acid after infection with bacteriophage T2." *Virology* 2: 149–61.

Watson, J. D. 1963. "The involvement of RNA in the synthesis of proteins." *Science* 140: 17–26.

Zallen, D. T. 1996. "Redrawing the boundaries of molecular biology: The case of photosynthesis." In *The Philosophy and History of Molecular Biology: New Perspectives*, ed. S. Sarkar. Dordrecht, Holland: Kluwer, 47–65.

Zallen, D. T., and R. M. Burian. 1992. "On the beginnings of somatic cell hybridization: Boris Ephrussi and chromosome transplantation." *Genetics* 132: 1–8.

Zamecnik, P. C. 1960. "Historical and current aspects of the problem of protein synthesis." *Harvey Lectures* 54: 265–81.

Zamecnik, P. C. 1969. "An historical account of protein synthesis, with current overtones – A personal view." *Cold Spring Harbor Symposium on Quantitative Biology* 34: 1–16.

Zamecnik, P. C. 1979. "Historical aspects of protein synthesis." In *The Origins of Modern Biochemistry: A Retrospect on Proteins*, eds. P. R. Srinivasan, J. S. Fruton, and J. T. Edsall. New York: New York Academy of Sciences, 269–301.

# 9

# Too Many Kinds of Genes?

## Some Problems Posed by Discontinuities in Gene Concepts and the Continuity of the Genetic Material (1995)[1]

Broadly speaking, there are two kinds of gene concepts. In this brief chapter, I offer a modest account of each. I seek to show that both are legitimate and that it is necessary to understand their interplay to understand the history of genetics and a number of current issues in genetics. I argue that the first kind of concept makes sense of the conceptual continuities in the history of genetics but yields concepts that are too generic or schematic to specify adequately what is referred to by "the" gene concept and allied concepts. I show that we cannot do without such generic or schematic concepts. Indeed, without schematic concept(s) of the gene, there would be no such discipline as genetics; however, without supplementation by more specific gene concepts, the schematic concepts do not suffice for specifying the reference of the term "gene" – indeed, they do not specify what genes are well enough to ensure that the term refers successfully at all. In less philosophical language, these schematic concepts are impotent to specify exactly what we are talking about when we talk about genes.

The second kind of gene concept, in contrast, yields specific gene concepts but does so at the price of conceptual discontinuity. I argue that if one restricts oneself to the series of discontinuous gene concepts, the findings of molecular genetics favor abandoning a univocal and specific concept of the gene altogether in favor of a pair of concepts: the concept of genetic material plus that of the expression of genetic information. I even suggest that, without the schematic concepts, molecular genetics would be well served by abandoning specific gene concepts with a concept of the genetic material. As some other contributors to the workshop in which this chapter was originally presented

[1] This text is revised from a paper presented to a workshop on gene concepts at the Max Planck Institute for the History of Science in Berlin in 1995. I am grateful to the discussants at that meeting for criticisms and suggestions. The content of the chapter has not been adjusted to reflect subsequent developments.

have argued,[2] the information content of the genetic material is extremely dependent on the cellular or subcellular context in which it is expressed. This provides one of the rationales for suggesting that molecular biologists could abandon specific concepts of the gene – deploying, instead, concepts focusing on the continuous genetic material and the controls governing what is still called gene expression.

## SCHEMATIC GENE CONCEPTS

Any science that seeks to locate hidden causes of some spatiotemporally delimited class of phenomena must use indefinite descriptions.[3] These are descriptions that leave the exact referent of a term open. An example would be a Mendelian description like *the factor, whatever it is, in the germ cells of these peas that causes them to produce plants that are shorter than the tall plants produced from peas from the same pod.* Such specifications are indefinite in not adequately specifying what the causal factor in question is or even what category or sort of thing or process the factor is. Indefinite descriptions can genuinely refer to entities as, indeed, the example I just gave – used in the right circumstances – does, but they can also be associated with seriously false descriptions or commitments. This is illustrated by the view, common before the middle of the twentieth century, that Mendelian factors (or genes) are composed of proteins. Mendelian genetics, taken strictly (i.e., without commitment to the localization of genes on chromosomes), used gene concepts based on very open-ended indefinite descriptions of exactly the form illustrated previously.[4]

[2] After a follow-up workshop at the Max Planck Institute, most of the papers alluded to were published in Beurton, Falk, and Rheinberger (2000). The papers by Falk, Fogle, Gifford, Gilbert, Holmes, Rheinberger, and Schwartz are particularly relevant to this chapter.

[3] I first learned the importance of indefinite descriptions in my graduate studies with Wilfrid Sellars. The concept of the reference potential of a concept, deployed in Chapter 7, is – in some respects – a development of one aspect of the indefinite reference of schematic gene concepts. For amplification of my views along these lines, see the treatment of referentially open concepts in Burian (2000, especially Sections 3 and 4).

[4] Johannsen's 1909 attempt at an a-theoretical definition illustrates the point precisely. In Carlson's translation (1966, pp. 20–2): "The word 'gene' is completely free from any hypotheses; it expresses only the evident fact that, in any case, many characteristics of the organism are specified in the gametes by means of special conditions, foundations, and determiners which are present in unique, separate, and thereby independent ways – in short, precisely what we wish to call genes." Genes are thus the differences, whatever they may be, between gametes that cause organisms to have the potential for revealing patent, independently heritable traits. Darden (1991) amplifies this point usefully in firmly separating Mendelian genetics as developed after the "rediscovery" of 1900 from the chromosomal theory developed by the Morgan group and others.

I call concepts like that of a gene thus understood *referentially indefinite causal (or functional) concepts*. In particular, the means of identifying genes just illustrated is indefinite, but the identification accomplished in terms of a two-part functional description. The first part specifies a difference in the phenotype of the organism bearing a gene (tall versus short); the second requires a pattern of transmission of the factor(s) responsible for the change. (One can distinguish different genes affecting, say, a plant's height or flower color by their behavior in breeding experiments, by whether they "Mendelize" or follow some recognizable variant of classic Mendelian patterns of inheritance.)

Here is a schematic formulation of a referentially indefinite functional gene concept: *A gene for trait X is any stably inherited factor that causes an organism [or certain cells of the organism], given the rest of what it has in common with conspecifics, to have the potential for manifesting X, where X will (or can be made to) appear under the appropriate developmental plus environmental circumstances.*[5] Distinct genes for X may exist and may be discriminated from each other either by specific differences in the phenotypes they cause or by demonstrating that they can be inherited independently of each other. Stadler (1954) used the label "the operational gene" for genes, thus indefinitely described. Two points were involved: First, there were competing theories, between which no decision was possible, of the constitution of operationally delimited genes. Second, breeding procedures allowed workers to distinguish between distinct genes with otherwise identical phenotypic effects.

Such concepts imply no direct claims about what genes *are* (e.g., what they are made of) or even whether they are chemical substances or stable harmonic resonances, which seems to be what Bateson thought they might be (Bateson 1913, Chapter 2; see Chapter 7, this volume). Without independent knowledge of gene structure or composition, then, these concepts do not provide a fully adequate way of individuating genes. (For that reason, Stadler spoke pessimistically of our inability to resolve questions about "the

[5] It is important to note that as we develop an account of the relevant causal chains, we may come to adjust what we count as a trait or, at least, what we count as a trait caused in a particular, stably inherited manner. Think, for instance, of the multiplication of distinct disease entities (e.g., distinct cancers formerly classified as a single disease) as we learned to distinguish different pathways by which the regulatory apparatus of the relevant cell types can be disrupted so as to yield phenotypically similar outcomes. It is also worth recognizing that schematic definitions may require specification in a great variety of ways. Thus, the specification of 'modifier genes' and 'regulatory genes' is often relative to a specific gene or control pathway carried by some, but not all conspecifics. Thus, the specification of the different alleles of a modifier gene may depend not just on differences caused by alternative alleles, but on differences caused by the alternative alleles when they are present in specific genetic backgrounds or when particular developmental contingencies – not experienced by all conspecifics – have produced certain features in the organisms carrying those alleles.

hypothetical gene" in distinction to the operational gene.) If no information about structure or composition is built into the gene concept, it is not possible to count genes in a stably satisfactory way. This helps make sense of the fact that the chromosome theory – or something like it – was flatly needed to complement or complete Mendelian genetics, and it helps explain part of what is accomplished by the specification of genes as composed of DNA and RNA. But, once such additional information is built into the concept of the gene, the theoretical presuppositions of gene concepts are radically strengthened – and, for most of the history of genetics, the presuppositions involved have been substantially false.[6]

One can view the history of genetics as involving, among other things, a series of attempts to obtain experimentally and conceptually sound ways of filling in indefinite descriptions of genes like those with which genetics began. What *should* count as a gene, given the indefinite starting point, depends on the specific traits or functions examined and the patterns of inheritance that they exhibit. It also depends on larger commitments as well, such as the means we employ in determining that something (e.g., a particular sequence of nucleotides), in context, is causally responsible for the trait differences in question. It depends, further, on the restrictions we place *in context* on the ascription of causal responsibility. In the century or so with which we are concerned, it has been at various times stoutly affirmed and stoutly denied that to count as a gene an entity had to be on or to be a part of a chromosome, or composed of protein, or composed of nucleic acid, and so on. In general, there is no adequate way of telling when such claims were intended as conceptual and when they were intended as factual. For this and other reasons, to make sense of the history of genetics we need to understand that *when such commitments had conceptual force, there was always a pathway of retreat*

[6] This claim is, of course, contentious, but I believe it is correct. Consider the sorts of substantially false conceptual commitments that have commonly been made: Genes are discrete particles, genes are composed of proteins, they are located only on chromosomes, they are linearly contiguous, they are non overlapping, and so on. When one claims that such commitments of detail have been built into gene concepts and are substantially false, one does not imply that genetics is based on fundamental mistakes. The ability to retreat to more generic concepts and the associated definitional muddiness, to which Rheinberger (2000) refers, allow for substantial conceptual falsity to coexist with a fundamentally sound theory or, better, a fundamentally sound program of research or research tradition yielding a sequence – or expanding family – of improving theories. These considerations support one of Rheinberger's contentions and work against another. In contrast to Rheinberger, I hold that we need concepts far more specific and less "dirty" than is justified by available evidence. He is correct, however, that such metalevel concepts as *research programme* or *research tradition* also must be open textured in much the same way that the concept of the gene is if *our* account of the history and conceptual structure of genetics is to be workable.

*open*. The underlying concepts to which people retreated when necessary were referentially indefinite functional concepts.

It should be clear that indefinite descriptions of genes, even when conjoined with massive sets of experimental results, are not sufficient to specify exactly to what terms like "gene" or "gene for X" and their cognates refer. One thing that is often meant by a (or "the") theory of the gene is the theory-based specification of what it is that goes into individuating genes beyond the indefinite descriptions plus sheer experimental findings. A great deal is involved here. Among them, for example, I include abstract principles for the delimitation of causes; the delimitation of the biological functions to be examined (cf. visible phenotypes versus protein synthesis); and commitments about the material composition, structure, or location of genes that constrain the concept of a gene and the possible referents of that concept. To understand the historical continuities that make genetics into a discipline and give geneticists a series of problematics on which to work, it is necessary to recognize not only this role of referentially indefinite concepts but also that referentially definite concepts (or, at least, referentially more definite concepts) are needed to specify *what genes are* and to develop means of testing the principal claims made about them: claims about how to individuate them, how they act, and so on. The need to answer such questions has had considerable impact on the character of theory in genetics. Indeed, the failure to develop globally satisfactory definite descriptions of genes is part of what moves me to suggest the need for conceptual reform in molecular biology.

## DISCONTINUOUS GENE CONCEPTS

More specific concepts of the gene, though they may still allow further specification, are committal – at least to some degree – about the structure or location of genes. What is typically required is a mixed mode of identification in terms of both structure and function. When such definite concepts embody false presuppositions, they may – if taken literally – turn out not to refer to anything (e.g., when they make the mistaken commitment that genes are composed of proteins) or they may apply to a subclass of the entities currently considered to be genes in molecular biology (as do those gene concepts that require genes to be composed of DNA, which miss the genes of RNA viruses).

It is always possible to retreat to a less definite description of genes and to set as a constraint on successful use of the terms in question that they refer to a causal factor contributing to the occurrence of a well-specified phenomenon. Of course, they might then end up referring to an integron (see Rheinberger 2000) and not DNA or RNA as such at all. Thus, it is (nearly) always possible to retreat from false presuppositions so that it is clear that the claims of scientists who employed those presuppositions made good sense (see Chapter 7, plus Burian, Richardson, and Van der Steen 1996; Kitcher 1978, 1982). But, it is also true that to individuate genes one must specify, among the thicket of factors contributing causally to any functional state, the substrates out of which genes are built and the structures that can count as relevant causes (and thus deserve to be identified as genes). Note that for this class of gene concepts, the choice of a phenotype is crucial in determining what counts as a gene: When the phenotype is an amino-acid sequence, genes will be individuated differently than when the phenotype is something like the suppression of the expression of certain other genes. And it will continue to be the case that biologists with different interests will seek genes for phenotypes of different sorts. Thus, one cannot escape the recognition that there are sharp discontinuities in the history of genetics – discontinuities that cannot be bridged directly ("genes must be composed of protein" versus "genes must be composed of nucleic acids"). Nonetheless, such differences can be bridged via a retreat to less definite descriptions.

Once this point is granted, it is clear that the findings of molecular biology, some of which I allude to briefly in the next section, are readily interpreted so as to call into question whether genes are particulate without preventing those of us who deny that they are particulate from referring to the same things that our forefathers in Morgan's and Bateson's groups did when they used terminology committed to particulate genes and dynamic equilibria, respectively. Indeed, in light of the treatment of concepts already given, I suggest that the findings of molecular biology allow one to challenge the claims that the terminology of genes is well defined and that it picks out a well-delimited group of entities. Given the range of functions for which we seek genes, one may even doubt whether all the gene-like causes are restricted to nucleic acids (cf. prions). However, I set that issue aside so that we may deal with the question of whether we have a good way of settling which parts of which DNA and RNA molecules ought to be considered to be genes in light of contemporary knowledge. To this question I believe there is no systematically satisfactory answer. The best answer in a given case depends on our purposes and on the schemes of classification we employ, both of the functions that may be caused genetically and of nucleic-acid molecules and their parts.

## CONTINUITIES IN THE GENETIC MATERIAL

Within rather broad limits, we are free to use terminology as we choose, although it pays to be reasonably clear about our usage and not to cause needless confusion by using preempted terms in ways that conflict with common usage. The term "gene" in molecular biology is a genuine accordion term – its expansion and contraction make for a lot of semantic music and allied quibbling. But, the arguments involved are not always empty semantic quibbles for they turn on the inclusion or exclusion of a number of genetic functions performed by nucleic acids that do not fit any of the standard structural constraints on genes. Underlying the different terminologies are serious disagreements about the status of parts of nucleic-acid molecules that behave or are treated in different ways in different cellular contexts and at different phases of ontogeny. I cannot pursue these issues very far here (see Chapter 12 for further discussion), but will take up a few matters briefly. For convenience, I work with one of the broader gene definitions (specifically, of eucaryotic genes) that I have encountered (Singer and Berg 1991).[7]

> We define a [eucaryotic] gene as a combination of DNA segments that together comprise an expressible unit, a unit that results in the formation of a specific functional gene product that may be either an RNA molecule or a polypeptide. The DNA segments that define the gene include the following:
>
> 1. The transcription unit refers to the contiguous stretch of DNA that encodes the sequence in the primary transcript; this includes (a) the coding sequence of either the mature RNA or protein product, (b) the introns, and (c) the 5′ leader and 3′ trailer sequences that appear in mature mRNAs as well as the spacer sequences that are removed during the processing of primary transcripts of RNA coding genes.
> 2. The minimal sequences needed to initiate correct transcription (the **promotor**) and to create the proper 3′ terminus of the mature RNA.
> 3. The sequence elements that regulate the rate of transcription initiation: this includes sequences responsible for the inducibility and repression of transcription and the cell, tissue, and temporal specificity of transcription. These regions are so varied in their structure, position, and function as to defy a simple inclusive name. Among them are **enhancers** and **silencers**, sequences that influence transcription initiation from a distance irrespective of their orientation relative to the transcription start site (pp. 461–2; see also pp. 435 ff. and 457 ff.).

[7] Chapters 11 and 12 contain illustrations that will help the reader unfamiliar with the technical terminology to understand Singer and Berg's text.

This definition includes a great deal that others would exclude. A more orthodox definition, like that of Goodenough and Levine (1974, p. 291), would restrict the gene to those nucleotides which, "when transcribed, will produce a biologically active nucleic acid," thus excluding promotor sites, enhancers, silencers, introns, and the like. But no matter: On either definition, most eucaryotic genes are discontinuous stretches of continuous DNA because introns are excised from biologically active RNAs. Worse yet, in many eucaryotes and quite a few procaryotes, chain termination is dependent on physiological circumstances and/or is developmentally regulated. This means that the size of a gene – or what parts of the DNA of a multigene family function as genes rather than counting as pseudogenes – depends on physiological circumstances or developmental stage. Worse yet are the cases in which there is partly programmed, partly random gene shuffling during ontogeny or in response to SOS signals. I cite mammalian immune systems and the rec-A system of *E. coli* and related systems in other bacteria as examples. Such shuffling of the genetic material means that the genetic contents of a zygote (i.e., a fertilized egg) are not preserved in certain somatic-cell lineages or even (in some cases) passed along to the offspring of fissioning bacteria. The dynamism of the genome is of great importance for the definitional and conceptual issues that belong at the heart of this chapter.[8]

It might be thought that this argument can easily be interpreted as a trivial semantic argument about how we should define terms and not as an argument bearing on how we should think about genes in light of the findings of molecular biology. The purpose of the next argument is to show that the issues just raised are not merely semantic in this pejorative sense but also seriously affect our interpretation of the history of genetics and impinge on how biologists should be thinking at this point. The moral that I wish to draw is that we should think not of "the molecular biology of the gene" (to use a well-known title [Watson 1965]) but of the "molecular biology of the genetic material." (I have borrowed this phrase from somewhere but I am no longer quite sure where – I believe I owe it to Dan Hartl.)

The argument is centrally concerned with the continuity of the genetic material; the intermediate conclusion that I draw is this: An examination of intrinsic features of RNA or DNA is not sufficient for the task of delimiting which parts of these molecules or which molecular structures should count as genes. The principal reason for this is simple: context dependence.[9] Ask

[8] For recent reviews providing some details and amplifying the importance of such issues, see Fogle (2000) and Portin (2002).

[9] This is also the reason for which the genetic code cannot be determined (or determined up to permutations) by an examination of the structure of DNA or mRNA molecules alone. In different

what bits of DNA or RNA perform functions that we wish to identify as genetic. It takes an enormous amount of machinery for genes to be expressed. There is a huge number of processing steps, nearly any one of which, in context, can affect the times and places at which informational molecules yield products – and just which products they yield. It was known as early as 1987 that the translational apparatus alone requires some 200 macromolecules (Freifelder 1987, p. 367)! Corresponding to the richness and variability of the mechanisms involved is the richness of the alternative results (even at the molecular level) when a given stretch of nucleic acid is transcribed or enters into an interaction of some sort. The answer to the question, *Which stretches of nucleic acid should count as genes?*, depends not only on the functions and the sequence of nucleotides that we have chosen to examine, but also on the particular machinery present in particular cells or compartments within cells, for it is the latter that determines which parts of the signal remain intact and are contiguously read out and what the molecular results of the network of interactions involved turn out to be.[10]

As is generally known, there is cellular machinery that determines which stretches of DNA are accessible to RNA polymerases, where it is that the RNA polymerases get stopped or knocked off the DNA (both dependent, for a given stretch of DNA, on physiological conditions), and how the resulting RNA is processed: immediately in procaryotes and before it can get through the nuclear membrane in eucaryotes. It is worth recalling at this point that, in eucaryotes, most genes are processed in such a way that the material corresponding to introns must be snipped out of the RNA molecule to allow the transcript to get through the nuclear membrane. At least occasionally, some of the material thus snipped out is, in turn, translated to yield a functional polypeptide (or is functional in some other way) so that it is natural to talk of one gene embedded inside another.[11] There may still be further post-transcriptional processing of mRNA[12] and, at that, the precise polypeptide sequence the RNA yields is still a function of the tRNAs in the relevant

cellular context (e.g., nucleus versus mitochondria; some species of organisms versus others), there are sometimes regular differences in the tRNAs. Thus, in a few cases, the same codon in different contexts codes for a different amino acid or for a stop signal instead of an amino acid.

[10] For some amplification concerning the points raised in this and the next paragraphs, see Chapter 12 and Keller (2000, 2003).

[11] A brief technical description of such a case is given by Singer and Berg (1991, pp. 705–6) for introns in the mitochondria of yeast.

[12] Alternative splicing is just one of many relevant post-transcriptional phenomena that are relevant here. Gilbert (2000) and Singer and Berg (1991, pp. 578 ff.) provide helpful accounts of alternative splicing and other technicalities discussed later. This phenomenon again demonstrates the impossibility of employing the intrinsic features of the DNA or RNA alone to determine

cytoplasmic location. Further, post-translational processing of proteins is, at least in some cases, critical to whether the product that results, in fact, enters into a final product that plays a functional role. Complications of the sort involved in the rec-A and immune system cases go on and on in many more ways than can be discussed here.

Perhaps a quasi-example will make the point clearer. Consider an ORF (i.e., a signal that opens a reading frame on a stretch of DNA), located by appropriate molecular techniques. Does the ORF mark the beginning of, or even delimit, a gene? The answer, so far as there is one, depends on the physiological context, the alternative splicing and readout controls present in the relevant cell (for the stop signals are different in mitochondria than in the nucleus), the tRNAs present in the immediate context, and so on and on. Often enough, a single ORF begins a transcript that contains multiple genes.[13] My conclusion is that even when one works at the molecular level, what counts as a gene is thoroughly context-dependent.

The arguments just presented interact constructively. Conceptually speaking, what counts as a gene depends on what one chooses as a phenotype. What one may choose as a phenotype, however, is somewhat constrained by what we learn about genes. Factually speaking, how one may delimit a gene at the molecular level depends on the entire system for processing DNA, RNA, the translation of processed RNA into protein, and also post-translational processing. As a result, the task of delimiting genes contains an inextricable mixture of conceptual and factual elements. To be sure, the "lowest level" (i.e., the molecular level), though it is most distant from naive observation, brings the argument closer to a context-fixed factual basis than the others. But, the price for this is that one must deal with the interactions of all of the relevant macromolecules within their physiological setting. This has the consequence that precise definitions of genes must be abandoned, for there are simply too many kinds of genes, delimited in too many ways. Taken in combination, these arguments combine to provide powerful support for the principal contention of this chapter: namely, that when we reach full molecular detail, we are better off to abandon specific gene concepts and to adopt, instead, a molecular biology of the genetic material.

which stretches of a DNA or RNA molecule produce "biologically active RNA." For further explanation of many of these issues, see Chapter 12.

[13] Chapter 5 of Gilbert (2000), which covers differential gene expression, includes useful reviews of differential RNA processing (pp. 130–3) and of (contextually variable) translational and post-translational controls of the end products of the expression of nucleotides sequences (pp. 134–6).

POSTSCRIPT 2003

Since 1995, much new attention has been given to the issues discussed in this chapter (e.g., Beurton, Falk, and Rheinberger 2000; Dietrich 2000; Griffiths and Neumann-Held 1999; Kay 2000; Keller 2000; Morange 1996, 2000, 2001; Moss 2001, 2003; Neumann-Held 2001; Portin 2002; Sarkar 1998; Snyder and Gerstein 2003; Waters 2000). One line of work is of special interest for the position staked out previously – namely, attempts to provide intrinsically molecular concepts of the gene. An important effort in this direction is Lenny Moss's *What Genes Can't Do* (Moss 2003). (For a contrasting approach to this problem, see Waters 1992, 1994, 2000.) Moss distinguishes sharply between two types of gene concepts, labeled *gene-P* and *gene-D*. Although the label gene-P is meant to capture the connection between preformationism and genes that somehow determine a phenotype, a gene-P is defined as a gene *for* a phenotype (i.e., one that is identified by its causal link to that phenotype) (see Moss 2003, p. 45). In contrast, a gene-D (the "D" indicates that the gene is interpreted as a developmental resource) is defined by its molecular sequence (i.e., intrinsically, without reference to what it produces). Moss rightly insists (as was argued previously) that a nucleotide sequence may enter into many different interactions and may be processed so that the products it yields have many different structures occurring in many different tissues. Similar things may be said for noncoding nucleotide sequences and the reactions that they affect. Accordingly, it is simply incorrect to identify molecular sequences in terms of particular effects. No gene-D is properly understood as a gene for X, where X stands for a phenotype or a function; the effects of a gene-D depend on the biological context and (often) on the history of the organism. Hence, the effects of a gene-D are "*indeterminate* with respect to phenotype" (Moss 2003, p. 45).

This point about nucleotide sequences and the indirectness of their relationship to phenotypes is entirely correct. But I am skeptical of Moss's deployment of the terminology of genes-D. The problem is how one delimits one gene-D from another. Not all nucleotide sequences should count as genes. Some short nucleotide sequences are repeated millions of times within the genome. Should each arbitrary length of such a sequence count as a distinct gene? For good reasons, even when one is working at the molecular level, it is often desirable to identify distinct nucleotide sequences as instances of the same gene – for example, in numerous contexts in which the relation between a gene and amino-acid sequences is at stake, synonymous substations are counted as alterations that do not change the identity of the gene, even at the molecular level. Moss would probably consider this a confusion of gene-P interpretations of the gene with gene-D interpretations of the gene. I consider it evidence that *even at the molecular level, functional criteria of delimitation are built into gene concepts*. The issues here obviously ramify far beyond this immediate, partly linguistic, partly conceptual point. Moss's insistence that we take seriously the idea of a sequence-defined or sequence-delimited concept of the gene is salutary. The issue between us (if it turns out not to be a confusion on my part) is over my claim that we need to restrict sequence-based definitions with further (functional) criteria to save the gene concept from picking out any and all arbitrary sequences. If I am right (and Moss may concur with this claim), the result is that *the context-dependence of the effects of nucleotide sequences entails that what a sequence-defined gene does cannot be understood except by placing it in the context of the higher order organization of the particular organisms in which it is located and in the particular environments in which those organisms live*. This argument provides a synopsis of the one strand of support for the claim that the science of genetics has argued itself out of the most stringent version of reductionism.

## REFERENCES

Bateson, W. 1913. *Problems of Genetics*. New Haven: Yale University Press.

Beurton, P., R. Falk, and H.-J. Rheinberger, eds. 2000. *The Concept of the Gene in Development and Evolution: Historical and Epistemological Perspectives*. Cambridge and New York: Cambridge University Press.

Burian, R. M. 2000. "On the internal dynamics of Mendelian genetics." *Comptes rendus de l'Académie des Sciences, Paris. Série III, Sciences de la Vie/Life Sciences* 323, no. 12: 1127–37.

Burian, R. M., R. C. Richardson, and W. J. Van der Steen. 1996. "Against generality: Meaning in genetics and philosophy." *Studies in History and Philosophy of Science* 27: 1–29.

Carlson, E. A. 1966. *The Gene: A Critical History*. Philadelphia and London: W. B. Saunders.

Darden, L. 1991. *Theory Change in Science: Strategies from Mendelian Genetics*. New York: Oxford University Press.

Dietrich, M. R. 2000. "The problem of the gene." *Comptes rendus de l'Académie des Sciences. Série III, Sciences de la Vie* 323, no. 12: 1139–46.

Fogle, T. 2000. "The dissolution of protein-coding genes in molecular biology." In *The Concept of the Gene in Development and Evolution: Historical and Epistemological Perspectives*, eds. P. Beurton, R. Falk, and H.-J. Rheinberger. Cambridge and New York: Cambridge University Press, 3–25.

Freifelder, D. 1987. *Molecular Biology*. Boston: Jones and Bartlett.

Gilbert, S. F. 2000. *Developmental Biology*, 6th ed. Sunderland, MA: Sinauer.

Goodenough, U., and R. P. Levine. 1974. *Genetics*. New York: Holt, Rinehart, and Winston.

Griffiths, P. E., and E. M. Neumann-Held. 1999. "The many faces of the gene." *BioScience* 49: 656–62.

Johannsen, W. 1909. *Elemente der Exakten Erblichkeitslehre*. Jena: G. Fischer.

Kay, L. E. 2000. *Who Wrote the Book of Life: A History of the Genetic Code*. Stanford, CA: Stanford University Press.

Keller, E. F. 2000. *The Century of the Gene*. Cambridge, MA: Harvard University Press.

Keller, E. F. 2003. *Making Sense of Life*. Cambridge, MA: Harvard University Press.

Kitcher, P. 1978. "Theories, theorists, and theoretical change." *Philosophical Review* 87: 519–47.

Kitcher, P. 1982. "Genes." *British Journal for the Philosophy of Science* 33: 337–59.

Morange, M. 1996. "Construction of the developmental gene concept. The crucial years: 1960–1980." *Biologisches Zentralblatt* 115: 132–8.

Morange, M. 2000. "Gene function." *Comptes rendus de l'Académie des Sciences. Série III, Sciences de la Vie* 323, no. 12: 1147–53.

Morange, M. 2001. *The Misunderstood Gene*. Transl. M. Cobb. Cambridge, MA: Harvard University Press.

Moss, L. 2001. "Deconstructing the gene and reconstructing molecular developmental systems." In *Cycles of Contingency: Developmental Systems and Evolution*, eds. S. Oyama, P. E. Griffith, and R. D. Gray. Cambridge, MA: MIT Press, 85–97.

Moss, L. 2003. *What Genes Can't Do*. Cambridge, MA: MIT.

Neumann-Held, E. M. 2001. "Let's talk about genes: The process molecular gene concept and its context." In *Cycles of Contingency: Developmental Systems and Evolution*, eds. S. Oyama, P. E. Griffith, and R. D. Gray. Cambridge, MA: MIT Press, 69–84.

Portin, P. 2002. "Historical development of the concept of the gene." *Journal of Medicine and Philosophy* 27: 257–86.

Rheinberger, H.-J. 2000. "Gene concepts: Fragments from the perspective of molecular biology." In *The Concept of the Gene in Development and Evolution: Historical and Epistemological Perspectives*, eds. P. Beurton, R. Falk, and H.-J. Rheinberger. Cambridge and New York: Cambridge University Press, 219–39.

Sarkar, S. 1998. *Genetics and Reductionism*. New York and Cambridge: Cambridge University Press.

Singer, M., and P. Berg. 1991. *Genes and Genomes: A Changing Perspective*. Mill Valley, CA: University Science Books.

Snyder, M., and M. Gerstein. 2003. "Defining genes in the genomics era." *Science* 300: 258–60.

Stadler, L. J. 1954. "The gene." *Science* 120: 811–19.

Waters, C. K. 1990. "Why the anti-reductionist consensus won't survive: The case of classical Mendelian genetics." In *PSA 1990*, vol. 1, eds. A. Fine, M. Forbes, and L. Wessels. East Lansing: Philosophy of Science Association, pp. 125–39.

Waters, C. K. 1994. "Genes made molecular." *Philosophy of Science* 61: 163–85.

Waters, C. K. 2000. "Molecules made biological." *Revue Internationale de Philosophie* 54: 539–64.

Watson, J. D. 1965. *Molecular Biology of the Gene*. New York: Benjamin.

# IV

# Development

Part IV is concerned mainly to set up for further work on what should be one of the most important topics in the philosophy of biology for the next decade: the interpretation of new work on animal development. Chapter 10 is primarily historical; it concerns the long-term difficulties of integrating genetics and development and the sharp divisions that arose between those two disciplines. Chapters 11 and 12 seek to interpret recent work aimed at integrating development, evolution, and genetics of animals. This work belongs to the new interdisciplinary field of "evolutionary developmental biology," which will be the jumping-off point for much of this work (see also Arthur 1997; Burian et al. 2000; Gilbert 2003; Hall 1999, 2000; Hall and Olson 2003; Raff 1996, 2000). As should be clear, these chapters deal with issues that are not stably resolved; they attempt to provide an interpretative framework that will help with the ongoing struggles to resolve a large number of open questions.

Chapter 10, previously unpublished, is closely based on a talk I delivered in the fall of 1986 at the Department of History and Sociology of Science at the University of Pennsylvania. There has been a considerable amount of historical work on the conflicts between embryologists and geneticists since 1986, so the chapter is by no means groundbreaking, but the fundamental conflicts involved are still not nearly as well known as they should be. The central issue is this: If nearly all cells of an organism contain the same genes, what system of controls determines the patterned differentiation and movement of cells that is required to make animal bodies? It is hard to see how the specifications that determine the structure of the body and its orderly construction can be contained in the genes alone. But if the genes alone do not contain the "program" specifying the structure and properties of the organism (in recent terms: if the nucleotide sequence of the DNA of a fertilized egg does not, by itself, contain the information necessary to specify the structure of the organism that will result from normal development of that egg), we need to (1) rethink the prevalent genetic determinism and our use of the metaphor of a genetic program, and (2) work out in detail the system of controls that make ontogenesis as reliable as it is. Because the puzzles around this topic remain important today, I include extended quotations to show that similar issues have been actively entertained for more than a century. As the final two chapters show, the problems about the control of development raised by embryologists over the last century are beginning to be adequately addressed only now – and the results are startling (see the next paragraph). The scale of the enterprise required to deal with these issues is enormous; the two following chapters reveal a very small tip of a very large iceberg.

Chapter 11, originally presented in 1995 at an international congress on logic, methodology, and philosophy of science, is an early attempt to digest one of the most startling findings of evolutionary developmental biology. In that year, scientists from Walter Gehring's laboratory in Basel showed that activation of a gene taken from a mouse and transplanted into fruitflies (*Drosophila*) could initiate the cascade of events that makes *Drosophila* eyes. Even more interesting, like the homologous fruitfly gene, the activation of the mouse gene could make eyes on patches of larval tissue (imaginal disks) fated to become wings, legs, antennae, and so forth on the adult fly. This result, based on detailed molecular work and biotechnological tools, shows that controls that set off the eye-making cascade within one organism (here, a fruitfly) can be activated by the products of genes from widely separated organisms (e.g., mice). Chapter 11 discusses this work as a step in the ongoing rapprochement between developmental biologists and geneticists. It provides sufficient background to understand the outlines of the phenomena involved and then examines consequences of the findings for a number of important topics: Does the result that the genes whose products set off the eye-making cascade are homologous show that the *eyes* of mice and fruitflies are homologous, rather than analogous as was traditionally believed? Should the genes in question be considered "master control genes?" To what extent does the explanatory apparatus developed to handle cases like this rest on a kind of holism? For example, are the controls best understood as integrated mechanisms within integrated cells or subcellular systems (which is my view), or can they be better understood "from the genes up" (which appears to be the view of some of the people who did the experiments showing that the mouse gene can trigger the *Drosophila* eye-making cascade)? I argue that the biological findings yield strong arguments for multilevel causation. If this is correct, these findings undermine rather than support genetic determinism.

Chapter 12, previously unpublished, continues discussion of the findings of evolutionary developmental biology. Earlier versions were presented publicly in 2001 and 2002. In the chapter, I build up the necessary biological background required to grasp some key points regarding how gene expression is controlled. Then I outline some findings showing that the control systems governing segmentation in animals also rest on very deep homologies – indeed, that variants of the same complex control system for segmentation are deployed in all or virtually all segmented animals. It is now clear that various controls that determine the cascades of events that produce segments (and also some particular organs – recall the material in Chapter 11) are preserved across enormous evolutionary distances. I use these results to reinforce the idea that evolution works by "tinkering" – by swapping around integrated units as well as undergoing point mutations. This finding is critical for the argument that multilevel causation is involved in both development and evolution. For one thing, within the life of a single organism, the same genetic material is used at different stages of ontogeny and in different tissues in quite different ways. (In this chapter I explore some of the consequences this has for how we think about genes. It appears that many genes are composed of functional subunits that, on an evolutionary scale, are swapped from one gene to another.) For another thing, the idea of tinkering radically changes the combinatorics involved in evolutionary processes: The accumulation of mutations is not the only mechanism for putting variants together. And it forces a deep reconsideration of the ways in which animals are constructed; the units out of which they are built include processes for segmentation, for building particular organs, and the like. This sort of claim explains some of the excitement generated by the new work on development. Evolutionary developmental biology is at the beginning of a long road, but the journey on which it has embarked promises major and fascinating revisions in our understanding of evolution, animals, and ourselves.

## REFERENCES

Arthur, W. 1997. *The Origin of Animal Body Plans: A Study in Evolutionary Developmental Biology*. Cambridge: Cambridge University Press.

Burian, R. M., S. F. Gilbert, P. M. Mabee, and B. Swalla, eds. 2000. Symposium on "Evolutionary developmental biology: Paradigms, problems, and prospects." *American Zoologist* 40 (5): 711–831.

Gilbert, S. F. 2003. *Developmental Biology*, 7th ed. Sunderland, MA: Sinauer.

Hall, B. K. 1999. *Evolutionary Developmental Biology*. Dordrecht, Holland: Kluwer.

Hall, B. K. 2000. "Evo-devo or devo-evo – does it matter?" *Evolution and Development* 2: 177–8.

Hall, B. K., and W. M. Olson. 2003. *Keywords & Concepts in Evolutionary Developmental Biology*. Cambridge, MA: Harvard University Press.

Raff, R. A. 1996. *The Shape of Life: Genes, Development, and the Evolution of Animal Form*. Chicago: University of Chicago Press.

Raff, R. A. 2000. "Evo-devo: The evolution of a new discipline." *Nature Reviews Genetics* 1: 74–9.

# 10

# Lillie's Paradox – Or, Some Hazards of Cellular Geography[1]

This chapter was written for a broad audience, including biologists, historians of science, philosophers of science, and sociologists. The topic offers something of interest to specialists in each of these fields, for it provides materials for case studies of conflicts with interacting conceptual, disciplinary, and institutional components among embryologists, geneticists, and workers in allied disciplines over the course of about seventy-five years.

## THE PARADOX

I happily take the blame for the somewhat obscure label "Lillie's Paradox," which I first heard used by Jane Maienschein. The paradox, widely known in various versions since about 1900, came to stand for a central failure of the research programs of Mendelian genetics, especially for embryologists. The failure endured from the 1920s forward into at least the 1950s. Indeed, some of the difficulties involved are, arguably, not yet fully resolved.[2]

This term honors Frank Rattray Lillie, a slightly younger contemporary and friend of Thomas Hunt Morgan. Lillie, as we shall see, provided a pithy formulation of the paradox and argued that it meant that Morgan's theory of inheritance would be unable to provide a viable account of ontogenesis (i.e., the development of an organism from a fertilized egg to an adult).

The paradox is easily understood. It turns on the fact that in virtually all multicellular organisms, virtually all cells have an entire complement

[1] This chapter derives from a lecture presented at the Department of History and Sociology of Science at the University of Pennsylvania in November 1986. The text has been improved but the content remains fundamentally unchanged to provide a historical snapshot of the state of historical and biological knowledge as of that date.

[2] That is, as of 1986.

of chromosomes and, thus, if you believe the chromosome theory, virtually all cells of a higher organism have the same genes. If this is correct, the differences between nerve and muscle cells, between phloem and xylem, cannot be due to differences in the genes. But, if that is the case, something else must determine the fate of particular cells and control the development of nerves and muscles, bone and sinew, kidney and liver, leaves and petals. Embryologists often argued, although I am not sure whether Lillie did, that whatever that something else is, it must be inherited or largely inherited. Otherwise, development would not be regular and like would not beget like. In any case, if the Mendelian–chromosomal theory of inheritance is correct, it must at best be only a partial theory, for it leaves out of account the system of inheritance that controls development. That, presumably, must be studied by different techniques than those of Mendelian genetics – perhaps those of embryology.

Here is Lillie's formulation of the paradox, taken from "The Gene and the Ontogenetic Process," published in 1927:

> With reference to the processes of embryonic segregation [i.e., differentiation – RB], genetics is to a certain extent the victim of its own rigor. It is apparently not only sound, but apparently almost universally accepted genetic doctrine to-day that each cell receives the entire complex of genes. It would, therefore, appear to be self-contradictory to attempt to explain embryonic segregation by behavior of the genes which are *ex hyp.* the same in every cell (Lillie 1927, p. 365).

This argument purports to establish that classical genetics does not (and perhaps cannot) provide *the* fundamental theory of heredity. Genetics can provide an analysis of the transmission of determinants (to use an old-fashioned word) or genes, and those determinants may alter the final phenotype of the organism, but genetics cannot provide an account of whatever it is that controls development so that the determinants contribute to the formation of an ordered whole, an organism with thus-and-such a fundamental structure. Whatever it is that does that job must do so by acting in an as-yet unknown manner to produce the correct temporal sequence and spatial distribution of stuffs and cells so as to yield (for example) bone, muscle, liver, kidney, and nerves, all properly organized to yield an animal (and similarly for plants.).

Without a theory that can explain the mechanisms of ontogenesis, so the argument runs, any account of inheritance is radically incomplete. By locating genes on chromosomes, classical genetics became committed to the equality of the genic complement in every cell and was thus unable to provide a theory of ontogenesis. Accordingly, embryologists could argue that the equality of the genes in every cell type proved the necessity of providing an account of an inherited system of controls governing differentiation independent of the transmission of genes.

Not all embryologists accepted this account of what was missing, and embryology did not offer a satisfactory theory of this putative second system of inheritance in the 1920s and 1930s. Nonetheless, embryology was at least properly situated to provide such a theory or to bring forth the necessary preliminary empirical knowledge of the ways in which development and differentiation work. Accordingly, Lillie sums up his stand as follows:

> The present postulate of genetics is that the genes are always the same in a given individual, in whatever place, at whatever time, within the life history of the individual, except for the occurrence of mutations or abnormal disjunctions, to which the same principles then apply. The essential problem of development is precisely that differentiation in relation to space and time within the life-history of the individual which genetics appears implicitly to ignore.
>
> The progress of genetics and of physiology of development can only result in a sharper definition of the two fields, and any expectation of their reunion (in a Weismannian sense [that is, into a unified theory of heredity]) is in my opinion doomed to disappointment. Those who desire to make genetics the basis of physiology of development will have to explain how an unchanging complex can direct the course of an ordered developmental stream (Lillie 1927, p. 367).

On the whole, geneticists thought that the paradox posed no fundamental threat to the foundations of their discipline. The differences between their reactions and those of the embryologists (and many other biologists interested in development) is in this respect quite fascinating, the more so in light of the fact that around 1900, the problems of heredity and development were generally thought to be inseparable. In this chapter, I start in that era to set some of the central issues in context, then examine certain features of the gulf that had opened up between genetics and embryology by the 1930s, and, finally, address certain aspects of the partial resolution of Lillie's paradox achieved by the late 1950s – and the conditions that had to be met to achieve that resolution.

## STAGE SETTING

The conflict we are examining was institutionalized, in the United States at least, as a conflict between geneticists and embryologists or, later, developmental biologists. I argue, however, that its roots were not originally institutional or disciplinary. The issues in question took their specific form during the great transformation of biology into an experimental science at the end of the nineteenth century. This transformation – as Garland Allen, Scott Gilbert, and Jane Maienschein, among others, have argued – was shaped by an important series of dichotomies belonging to turn-of-the-century biology – dichotomies between preformation and epigenesis, structure and function, morphology and physiology, nucleus and cytoplasm, discontinuous and continuous variation, heredity and development.

Maienschein's work suggests a useful approach to the situation of experimental biologists interested in heredity and development around the turn of the last century (see, e.g., Maienschein 1981, 1983, 1986, 1987, 1991). A number of them – Conklin, His, Loeb, Morgan, Whitman, and Wilson are good examples – sought to achieve a *rapprochement* between the approaches taken by biologists whose predilections lay on opposite sides of the dichotomies on my list. Numerous attempts were made to reconcile disparate styles of work and the conflicting theoretical commitments with which they were allied. The issues in question centered on cytology, embryology, evolution, morphology, and theoretical analyses of heredity. As we will see, the rapprochement failed, and the most prominent parts of the work with which I am concerned were ultimately distributed into the disciplines of genetics and embryology.

I argue that, as things developed during the century, the distinction between nucleus and cytoplasm – what Ephrussi in the late 1950s called the "geographical distinction" – came to be an obstacle to an integrated understanding of heredity and development, an obstacle that had to be overcome before there could be any real hope of breaking down the conceptual barriers between genetics and developmental biology. One of the reasons for this is that cellular geography became connected with specific positions with respect to the dichotomies that I listed. A principal target of this chapter is to gain a better understanding of this fact.

## THE ISSUES

It is now widely recognized among historians of biology that the separation of studies of heredity into two distinct disciplines, which had the labels "genetics" and "embryology" by the late 1920s, occurred during the period between, roughly, Weismann's publication of *Das Keimplasma* (*The Germplasm*) in 1892 (1892, 1893) and Morgan's *The Theory of the Gene* in 1926. There has been considerable study of the formation of genetics as a separate discipline – although I think it has not been recognized how much the growth of genetics in the United States shaped the institutionalization of genetics in its early phases – and rather less investigation of the formation of twentieth-century embryology as a distinct discipline. I address three aspects of the separation of these two disciplines: conceptual issues, matters of experimental methodology, and sociological and institutional issues that came to be of pressing importance. An analysis of the interactions between the scientists involved or of the disciplinary dialectic that they produced must deal with all three of these, however sketchily.

I start with some conceptual points. The application of experimental technique to the study of heredity and development had its beginnings some ten or fifteen years before the turn of the last century. At the time, particularly in Germany, there were a number of self-conscious attempts to bring experimental techniques to bear in investigations concerning the interconnections among evolution, heredity, and development. The major figures hoped to provide an alternative to "merely descriptive" natural history and embryology and either to resolve the issues raised by such speculative theorizing as that of Haeckel and Darwin and their intellectual descendents or to set those issues aside. There was growing discontent with the inability of evolutionary theory to supply sound evidence for its speculative phylogenies and for the adaptive or nonadaptive scenarios that were being offered to account for evolutionary history. There was also strong interest in setting biology on a sound mechanistic or physicalistic foundation. Finally, there were all of the new discoveries regarding the nucleus and the internal structures of cells, hastened and facilitated by improvements in microscopy and, especially, in staining techniques. Accordingly, considerable pressure was generated to develop a suitable experimental methodology for investigating issues touching on heredity and to reformulate the problems involved in such a way as to allow the posing of experimentally resolvable questions. The test of explanatory adequacy in this new tradition would be demonstration of experimentally adequate mechanisms rather than explanatory consilience of speculative theories in the style of Darwin. For general reviews that introduce the relevant literature, see Maienschein (1990) and Olby (1990).

At the same time, it was impossible to escape the larger issues posed by the grand theories and findings of the nineteenth century – and the styles of work developed during that century. It was impossible for biologists with theoretical interests to avoid taking positions with respect to the conflict between old-fashioned morphology and embryology (with their *Baupläne* and *embranchements* of the animal kingdom) on the one hand and theories of evolution, whether Darwinian or not, on the other. It was impossible to avoid some sort of contact with the issue of preformation versus epigenesis in ontogeny. It was impossible to escape the questions posed by the discovery of the ubiquity of cells and nuclei and the growing knowledge of the dance of the chromosomes in mitosis and meiosis. And it was impossible for biologists with synthetic interests not to face up to issues regarding the production and the inheritance of variation.

Speaking loosely, the grand divide with which I am concerned is marked by the reactions of different investigators to the findings regarding the nucleus made from about 1885 on. Those who came to see the nucleus as insulated from traffic with the external world and to regard the seemingly equal contributions of egg and sperm nuclei to the properties of the progeny of their union as decisive – for example, Weismann and Roux (see Roux 1883, translated as Roux 1991) – tended toward some form of preformationism. After all, if the sperm – which contributed only its nucleus to the zygote – is responsible for half of the inherited characters of the organism, then the two nuclei together must be the source of *all* inherited characters. If the nucleus is insulated from events in other cells and insulated from the environment, inheritance of acquired characters must be impossible. And, if nuclear contents are fixed at the moment of the union between sperm and egg nuclei, the entire hereditary potential of the organism must be determined at that point. Furthermore, as it slowly became clear that – despite the exceptional behavior of the chromosomes of such organisms as *Ascaris*[3] – the chromosomes are constant and individual, definite structures became available to serve as the bearers of the determinants that governed the characters of the organism. Such considerations fostered an easy alliance between morphologically inclined microscopists and preformationists, between Weismannian evolutionists and advocates of discontinuous variation. (Incidentally, Weismann was one of those who used the term "evolution" in its old meaning of unfolding according to a predetermined plan, so that he concluded *Das Keimplasma* by remarking that ontogeny proceeds by evolution rather than epigenesis!)

On the other hand, many considerations from the study of development pointed in an opposite direction. Differentiation, it became clear, depended on the formation and localization of unlike substances in different regions of the protoplasm of the egg. To play any role in metabolism, differentiation, or regeneration, the nucleus had to be in a cytoplasm: It mattered which

[3] The phenomenon of chromosome diminution, or shattering, in somatic cells was discussed in numerous papers by Boveri (1887, 1892, 1902, 1910) and van Beneden (van Beneden and Neyt 1887).

cytoplasm it was in and, in at least some cases, where it was in the cytoplasm. Many texts make similar claims (e.g., Driesch 1894, Morgan 1910, Wilson 1896); Conklin, for one, put it this way in 1908:

> Neither the nucleus nor the cytoplasm can exist long independently of the other; differentiations are dependent upon the interaction of these two parts of the cell; the entire germ cell, and not merely the nucleus or cytoplasm is transformed in the embryo or larva; and it therefore seems necessary to conclude that both nucleus and cytoplasm are involved in the mechanism of heredity (Conklin 1908, p. 95).

Some biologists, including Conklin, took a much stronger stand in favor of epigenesis. In general, those whose work focused on events in the cytoplasm of the egg found asymmetries and local differentiations that, as cell lineage studies showed, were part of the sequence of events involved in determining how cells would differentiate and what their fate would be. After surveying a wide range of such evidence, Conklin reached the following italicized conclusion:

> *at the time of fertilization the hereditary potencies of the two germ cells are not equal, all the early development, including the polarity, symmetry, type of cleavage, and the relative positions and proportions of future organs being predetermined in the cytoplasm of the egg cell, while only the differentiations of later development are influenced by the sperm. In short, the egg cytoplasm fixes the type of development and the sperm and egg nuclei supply only the details* (Conklin 1908, p. 98).

The poles of the debate are now sketched in. Those who focused on the nucleus were at one extreme. They found morphological structures (the chromosomes), which they believed to be, or to carry, the determinants of heredity. These varied discontinuously, yielding what came to be known as mutations. Such views fall within a preformationist tradition. The biologists who focused on the physiology of development were at the other extreme. They found continuous variation in the cytoplasm, with regional differentiations affecting the properties of cells in various ways and sometimes determining the fate of the descendants of those cells. Most of them treated hereditary variation as continuous and held that there was at least some room for environmental influence on heredity, as shown by the consequences of alterations of egg cytoplasm. These views belong in an epigenesist tradition. In the middle were many who hoped to reconcile these two aspects of heredity.

## SEPARATION

Between 1890 and 1910, the term "heredity" underwent a transformation. The change of meaning is not easy to track because it took place in stages and because the transformation worked differently in different individuals, disciplines, and national traditions. In the case of Thomas Hunt Morgan, who, until the latter part of 1910, was a friend of the cytoplasm in the style of Conklin, the transition was abrupt. Here, for example, are the opening sentences of his last major anti-Mendelian piece, published in 1910:

> We have come to look upon the problem of heredity as identical with the problem of development. The word heredity stands for those properties of germ-cells that find their expression in the developing and developed organism. When we speak of the transmission of characters from parent to offspring, we are speaking metaphorically; for we now realize that it is not characters that are transmitted to the child from the body of the parent, but that the parent carries over the material common to both parent and offspring (Morgan 1910, p. 449).

This definition of heredity is the old one; it does not distinguish clearly between determination of a hereditary potential and differentiation or development of that potential. That distinction is precisely the one that was hammered out, at least in part, during the period in question, most especially in connection with the determination of sex. In fact, Morgan was a late convert; by 1910, cytological work had fairly well established that sex is determined in at least some cases by the chromosome complement and/or the determinants it carries. The determination of sex made it clear that one could learn *what* was transmitted (i.e., in one sense, inherited) and that what was transmitted (e.g., an X rather than a Y chromosome) then determined or, together with other factors, codetermined major properties of the organism (e.g., sex). Furthermore, understanding this did nothing to show *how* the properties or characters determined by the inherited constitution are brought about and yielded no explanation of the control of development. In short (and Morgan proclaimed as much as early as 1911), it is necessary to distinguish hereditary determination of (the potential for producing) characters from development or differentiation, the process by which those characters are realized. Here is how he formulated the point in 1917:

> The most common misunderstanding arises, I venture to think, from a confusion of the problem concerned with the sorting out of the hereditary materials (the genes) to the eggs and sperms, with the problems concerning the subsequent action of these genes in the development of the embryo (Morgan, 1917, p. 514).

It is readily apparent that investigation of heredity in the new sense calls for radically different methods than investigation of development. Not surprisingly, geneticists were keenly aware of this fact. Thus, as early as 1915, Morgan, Sturtevant, Muller, and Bridges defended the independence of their methodology from any considerations regarding development in their famous text, *The Mechanism of Mendelian Heredity*:

> It is sometimes said that our theories of heredity must remain superficial until we know something of the reactions that transform the egg into an adult. There can be no question of the paramount importance of finding out what takes place during development. The efforts of all students of experimental embryology have been directed for several years toward this goal. It may even be true that this information, when gained, may help us to a better understanding of the factorial [i.e., gene] theory – we can not tell; for a knowledge of the chemistry of all of the pigments in an animal or plant might still be very far removed from an understanding of the chemical constitution of the hereditary factors by whose activity the pigments are ultimately produced. However this may be, the far-reaching significance of Mendel's principles remains, and gives us a numerical basis for the study of heredity. Although Mendel's law does not explain the phenomena of development, and does not pretend to explain them, it stands as a scientific explanation of heredity, because it fulfils all the requirements of any causal explanation (Morgan, et al. 1915, pp. 226–7 and at pp. 280–1 in the 1922 rev. ed.).

An interesting feature of the institutionalization of biology in America during the first two decades of this century is the way in which institutionally independent situations were created for biologists who worked with the two different accounts of heredity. In good part, this was possible because of the expansion of biology during this period in America. The expansion made possible the formation of new institutions that fostered work in a large variety of styles, without excessive concern about making contact with the issues raised by those who worked in other traditions. In this regard, the situation in America is quite different than that in any of the European countries. Correspondingly, the separation of genetics from embryology was more severe in this country than it was in Europe.

In any event, genetics was institutionalized in the United States in ways that did not require geneticists to maintain contact with issues in evolution or development.[4] This may be seen in the formation and history of the genetics groups at such places as Cornell, Columbia, The Bussey Institution at Harvard, and the Cold Spring Harbor Laboratory of Experimental Evolution;

[4] Some starting points into the extensive literature on this topic are in Benson, Maienschein, and Rainger (1991) and Rainger, Benson, and Maienschein (1988).

in the development of genetic research in many of state agricultural stations, state land-grant colleges and universities, and private colleges; in the circumstances surrounding the founding of such journals as *The Journal of Heredity* and *Genetics* and the transformation of other journals such as *The American Naturalist*; and so on. By a variety of means, American genetics acquired an institutional base far stronger than any in Europe, one that fostered research into heredity in the new sense on a scale not possible elsewhere, one that removed the responsibility of facing the issue raised by such friends of the cytoplasm as Conklin and Lillie.

By the late 1920s, there was no hope of reconciliation between genetics and embryology on the basis of ongoing work of the types then common in the United States. The institutional separation of research in genetics and embryology was thorough and the programs of research of the principal figures across these disciplines were sufficiently divergent that there was little chance of bringing their work into intimate contact. Furthermore, the practical success and high reputation of genetic research left many embryologists feeling jealous and threatened. A major boundary – indeed, a political border as Allen (1974, 1986), Gilbert, (1988), and Sapp (1983, 1987) have shown – had arisen between the two disciplines, between students of the nucleus and students of the cytoplasm or the cell as a whole. Observers watching this boundary in the 1930s might notice some reconciliatory forays across the border, but they would also see a variety of breastworks going up and a number of potshots being taken.

## HOW SERIOUSLY SHOULD WE TAKE THE PARADOX?

There are a great many signs that the estrangement between embryology and genetics during the 1930s was serious. In the first instance, the conflict was over the degree to which the materials present in the nucleus of a fertilized egg were *by themselves* sufficient (or nearly sufficient) to account for the potentialities or the characters of the fully developed organism that would grow out of that egg. But, as I have been arguing, the conflict was in many ways a continuation of an older conflict between preformationists and epigenesists and was caught up in disagreements over the proper tools for examining such issues.

Lillie's Paradox was widely known (though not under that label) and widely thought by embryologists to be of great importance as an indicator of the limitations of genetics. For example, in France, where the problem of development was always influential in theories of heredity, the paradox was widely deployed as a central argument against the fundamentality of Mendelian inheritance. Lillie's formulation of the paradox had no particular influence there, but it is common to see the paradox used in support of an argument for there being a second – and far more fundamental – system of inheritance over and above the Mendelian one.[5]

The quotations in the appendix to this chapter show how easy it is to find versions of the paradox in the literature. A second glance at those quotations shows considerable variation in the conclusions drawn from the paradox. Let us ask exactly what inferences should be based on the paradox and whether it ought to have been taken so seriously.

One of the earliest statements of the paradox I have found was made by Driesch in 1894:

> Insofar as it carries a nucleus, every cell, during ontogenesis, carries the totality of all primordia; insofar as it contains a specific cytoplasmic cell body, it is specifically enabled by this to respond to specific effects only . . . When nuclear material is activated, then, under its guidance, the cytoplasm of the cell that had first influenced the nucleus is in turn itself changed, and thus the basis is established for a new elementary process, which itself is not only a result but also a cause (Driesch 1894).

Driesch's response to the uniformity of the contents of the nucleus, like that of Wilson in 1896, supposes that there is interaction between chromosomes and cytoplasm in such a way that the primordia – which, for present purposes can be lumped with Mendelian factors – are activated or otherwise regulated by the cytoplasm, initiating a series of reciprocal interactions that determine the course of determination and development. In contrast to Driesch, Wilson, a friend of the nucleus, goes on to say the following:

> In accepting this view we admit that the cytoplasm of the egg is, in a measure, the substratum of inheritance, but it is so only by virtue of its relation to the nucleus which is, so to speak, the ultimate court of appeal. The nucleus cannot operate without a cytoplasmic field in which its peculiar powers may come into play; but this field is created and moulded by itself. Both are necessary to *development*; the nucleus alone suffices for the *inheritance* of specific possibilities of development (Wilson 1896, p. 327).

Thus, before the turn of the last century, two general schemes that allowed a response to the paradox were in hand. Why were conceptual and experimental resolution so difficult? What further considerations gave the paradox its bite?

[5] On the French tradition, see Burian and Gayon (1999); Burian, Gayon, and Zallen (1988, 1991); and Gayon and Burian (2000).

I do not have a full answer to these questions, but it seems quite clear that the answers lie in the commitment of friends of the nucleus, including geneticists, to some version or other of preformationism and to certain hypotheses about the physiology of genes. The genes are what they are; they always act, and they must all act to make the organism. This illustrates my claim that the *geographically* distinct entities, nucleus and cytoplasm, became *conceptually* tied to philosophical stances regarding the manner in which organisms are formed. Many friends of the nucleus thought that there had to be some sense in which, once the nucleus was formed, the organism was all there, at least potentially, at least in its hereditary traits. The friends of the cytoplasm thought that no entities could determine the organism in that way, that the process of development that was revealed in their observations and experiments took precedence over, or dominated, any mere material particles or substances in determining what an organism would become. As the following quotations and those in the appendix show, commitments of these sorts can be found on all sides:

> If Mendelian characters are due to the presence or absence of a specific chromosome [or part of a chromosome], as Sutton's hypothesis assumes, how can we account for the fact that the tissues and organs of an animal differ from each other when they all contain the same chromosome complex . . . [W]e must be prepared to admit that the evidence is entirely in favor of the view that the differentiation of the body is due to other factors that modify the cells in one way or in another. This consideration is, to my mind, a convincing proof that we have to deal with two sets of factors – the common inheritance of all the cells to produce all the kinds of tissues and organs in the body, and the limitation of that property in the course of development. If the former is due to the chromosomes and the unspecialized parts of the cytoplasm, the latter may be due to the local changes that the relation of the parts to each other calls forth. It might even be argued that since in the development we find no evidence of a sorting out of the chromosomes that produce special parts, the individual chromosomes can not stand each as the representative of these parts, but rather that each part needs the entire set of chromosomes for its normal life (Morgan 1910, p. 477).
>
> As I have already pointed out, there is an interesting problem concerning the possible interaction between the chromatin of the cells and the protoplasm during development. The visible differentiation of the embryonic cells takes place in the protoplasm. The most common genetic assumption is that the genes remain the same throughout this time. It is, however, conceivable that the genes also are building up more and more, or are changing in some way, as development proceeds in response to that part of the protoplasm in which they come to lie, and that these changes have a reciprocal influence on the protoplasm. It may

be objected that this view is incompatible with the evidence that by changing the location of cells, as in grafting experiments and in regeneration, the cells may come to differentiate in another direction. But the objection is not so serious as it may appear if the basic constitution of the gene always remains the same, the postulated additions or changes in the genes being of the same order as those that take place in the protoplasm. If the latter can change its differentiation in a new environment without losing its fundamental properties, why may not the genes also? The question is clearly beyond the range of present evidence, but as a possibility it need not be rejected (Morgan 1934, p. 234).

Can we refute one or the other of these positions [that "the totality of the hereditary material is given in the genes" or that there are two systems of inheritance] by experiments at this time? Not by means of *classical Mendelism*. For in order to recognize or to localize genes, we need two organisms that differ in some particular traits, and even if we can trace certain fundamental properties (such as chromosome form or size) back to Mendelian genes, it appears to me to be completely impossible to remove the force of the dualists' arguments by analysis of hybrids. Johannsen put it like this: "... *By analysis of hybrids we have examined only the clothing*, the underlying organization remains unanalyzed. Whether we will ever be able to strip kinds of organisms [*Gattungen*, a word then used for races, species, and genera in German] of their superficial characteristics in such a way as to reveal the ultimate X of our formulas – a fundamental substance, something quite general and organic, something that, like a homozygote, is not accessible to Mendelian analysis – that remains an unanswerable question (Hertwig 1934, p. 428) [translated by RB, emphases in the original, references omitted].

Since cellular differentiation takes place in the cytoplasm, we are concerned here more directly with this constituent of the cell, bearing in mind, however, that the cytoplasm is accompanied and ultimately controlled by the genic complex in the chromosomes. Since the latter is presumably the same for all cells of the organism, differences between cells must arise through interaction between the constant genom [sic] and the locally variable cytoplasm in which they ultimately become visible (Harrison 1940, p. 93).

Now that the necessity of relating the data of genetics to embryology is generally recognized and the "Wanderlust" of geneticists is beginning to urge them in our direction, it may not be inappropriate to point out a danger in this threatened invasion.

The prestige of success enjoyed by the gene theory might easily become a hindrance to the understanding of development by directing our attention solely to the genom [sic], whereas cell movements, differentiation and in fact all developmental processes are actually effected by the cytoplasm. Already we

> have theories that refer the processes of development to genic action and regard the whole performance as no more than the realization of the potencies of the genes. Such theories are altogether too one-sided (Harrison 1937, p. 372).

The main point that I wish to make by means of this small selection of relevant quotations is that the paradox becomes decisive as soon as it is thought *both* that the nucleus contains all of the determinants for the hereditary properties of the organism *and* that the full complement of genes is always essentially the same in all cells. For if the nucleus or its contents act as a unit to achieve determination and the nuclei in all of the different cells of an organism are equivalent, then there is nothing left by reference to which one can account for the differences between the cells. Stated this way, it becomes obvious that the paradox represents a serious challenge to the research program of classical genetics, for that program *was* committed to the claims just made: The full complement of genes is present in virtually all cells; the nucleus contains, with but minor exceptions, all of the hereditary determinants of the organism; and all genes are involved in the formation of all properties of the organism.

## STEPS TOWARD A RESOLUTION

The problem of the control of development, posed in one form by Lillie's Paradox, is by no means wholly resolved today. Yet, the paradox, in the forms we have been examining, no longer poses a threat to contemporary genetics. I turn to aspects of the transition that tamed the paradox not yet widely examined in the literature, although some of them have been discussed in articles by Gilbert (1988, 1991) and Sapp (1987); see also Hamburger (1980) and Sapp (1986). Sapp examines various attempts to break the nuclear monopoly within genetics by finding factors, substances, or states in the cytoplasm that determine heredity. As he shows, difficulties of the type we have been discussing were employed to support a number of diffuse countercurrents within genetics, mostly in Europe and mostly outside the mainstream of the institutionalized discipline. By the 1950s, however, the situation had changed. Various workers around Tracy Sonneborn in this country, a number of French geneticists around Ephrussi and Lwoff, many of the people working on the genetics of single-celled organisms and phage, various Germans (some expatriate), and a considerable number of allies of the Morgan school were seriously investigating cytoplasmic phenomena.

I highlight here something that Sapp points out, but which I gloss somewhat differently than he does. This is the importance of the recognition, achieved in some circles during the late 1950s, that the geographical distinction between nucleus and cytoplasm was not the key to understanding control of differentiation and development but rather a red herring and a distraction. The critical point was an abstract one; it turned on developing a distinction between the genetic system and the allied epigenetic control systems that regulate the activity of genes, wherever they are located, and on working out the details of the interactions between genetic and epigenetic systems.[6]

There are far too many strands to the story of the transition between the two points of view for me to enter into serious detail. Instead, I point out a couple of markers that illustrate the character of the change and suggest where to turn to work out an account of it. The markers are, I think, more important than is widely known, but they do not contain the germ of the whole story, which involves elements from biochemistry, DNA research, embryology, cytology, mainstream Mendelian genetics, operations research, virology, and so on. Nonetheless, the work to which I call attention precedes by a few years the most frequently cited turning point in the analysis of gene regulation: namely, the Lwoff-Monod-Jacob-Elie-Wollman et al. work on the operon. What is involved is precisely the recognition that the geographical distinction between nucleus and cytoplasm worked a disservice in the attempt to understand hereditary control of development, especially insofar as it was allied to the sort of preformationist–epigenesist debate that has been our subject to this point. The catalyst (though, of course, it comes at the end of a long chain of preparatory studies) was an unpublished talk by David Nanney, a student of Sonneborn (a leading worker on paramecium genetics and a strong advocate of the importance of cytoplasmic genetic factors) at a meeting on Extrachromosomal Heredity held early in 1957 in France. Nanney's views were put forward in a widely influential article of Ephrussi's, "The cytoplasm and somatic cell variation" (Ephrussi 1958), delivered in 1957, and in Nanney's article, "Epigenetic control systems" (Nanney 1958). I cite Nanney at some length and then add a brief conclusion.

Nanney's article begins with a paragraph summarizing recent work supporting the ideas that DNA (or occasionally RNA) is the genetic material, that a single strand of DNA serves as the template for the reproduction of a complementary strand, and that certain regions in a strand of DNA (or its

[6] Two entirely different starting points for developing accounts of the interactions of genetic and epigenetic systems are found in Darlington's work on plant genetic systems, first synthesized in Darlington (1939) and in work on "infectious heredity" in bacteriophage and other microorganisms (e.g., Darlington 1944; Wollman 1928; Wollman and Wollman 1925). (Eugène Wollman, whose work is cited here, was the father of Elie Wollman, mentioned in the next paragraph of the text.)

complement) embody information in some way so as to specify the sequence of amino acids in corresponding proteins. He then writes:

> This view of nature of the genetic material . . . permits, moreover, a clearer conceptual distinction than has previously been possible between two types of cellular control systems. On the one hand, the maintenance of a "library of specificities," both expressed and unexpressed, is accomplished by a template replicating mechanism. On the other hand, auxiliary mechanisms with different principles of operation are involved in determining which specificities are to be expressed in any particular cell. Even without specifying precisely how these other mechanisms operate, the distinction between mechanisms involving template replication and "other mechanisms" is reasonably clear, even though both are involved in determining cellular characteristics. Difficulties arise, however, when one attempts to determine whether observed differences in cellular properties are due to differences in the "primary genetic material" or to differences in other cellular constituents. Some of the difficulties can be made apparent by setting forth certain general propositions related to the supplementary regulatory systems for which evidence is now available. To simplify the discussion of these two types of systems, they will be referred to as "genetic systems" and "epigenetic systems." The term "epigenetic" is chosen to emphasize the reliance of these systems on the genetic systems and to underscore their significance in developmental processes (Nanney 1958, p. 712).

Here are Nanney's five general propositions, with a small part of his amplifying text:

1. Cells with the Same Genetic Material May Manifest Different Phenotypes . . .
2. The Genetic Potentialities of a Cell Are Expressed in Integrated Patterns . . .
3. Particular Patterns of Expression Can Be Specifically Induced . . .
4. Epigenetic Systems Show a Wide Range of Stability Characteristics . . .

   These observations [of stable epigenetic homeostasis] create a real problem. One operational definition for "hereditary differences" has involved the indefinite perpetuation of cellular differences during growth in the same environment. Yet instances are known in which cellular differences may be maintained in the absence of detectable genetic or environmental differences. Hence the observation of indefinite persistence of differences does not distinguish persistent homeostasis due to DNA maintenance (genetic homeostasis) from persistent homeostasis due to epigenetic regulation (epigenetic homeostasis). Moreover, great difficulty is encountered in separating, on any conceptual basis, epigenetic systems controlling differences which persist indefinitely from systems maintaining differences for shorter periods of time, and these, in turn, from systems relating to differences

> which disappear immediately when cells are placed together in the same environment. "Cellular memory" is not an absolute attribute . . .
>
> 5. Some Epigenetic Devices May Be Localized in the Nucleus.
>
> Some attempts to characterize cellular regulatory agencies have placed considerable reliance on a geographical distinction; the genetic systems were considered to reside in the nucleus and on the chromosomes, to be stable and insulated from environmental alterations. The supplementary systems were thought to occupy the cytoplasm, to be more flexible and responsive to environmental alterations. While this distinction may have some general validity, its usefulness in particular cases may be slight. First, some of the systems of greatest interest are not amenable to the operations permitting a distinction between nuclear and cytoplasmic bases, i.e., breeding analysis or nuclear transplantation. Second, some genetic material occurs in the cytoplasm, although its common occurrence there is debatable. More serious are observations which suggest that some epigenetic control systems are located in the nucleus. The studies on nuclear differentiation in amphibian development provide perhaps the most dramatic single example of this evidence. Studies on the serotypes in *Salmonella*, however, go even further in suggesting that such control systems may even be localized on the "chromosomes" themselves. If such systems are so localized and particularly if they manifest considerable stability, they would behave in breeding analyses in a manner strictly comparable to genetic systems and would be indistinguishable from them on this basis alone . . .

In short, Nanney argues that

1. the distinction between nucleus and cytoplasm is largely irrelevant or only accidentally relevant;
2. the crucial distinction is between encoded information (contained in the genetic system) and the various systems for releasing or controlling the expression of that information; and
3. with the operational tools available in the late 1950s, one could not securely distinguish between changes of genetic content and long-term self-perpetuating changes in the state of a control system.

Although Nanney and Ephrussi were, perhaps, not the most important proponents of these three claims, and although the claims were supported by complex and broad sorts of evidence drawn from various investigations, the rapid acceptance of these or similar views among many geneticists and at least some developmental biologists in the late 1950s and early 1960s marks the end of the story to which this chapter is devoted.

## CONCLUDING REMARKS

Most biologists these days resist the intrusion of philosophers and other outsiders on their territory. This is understandable – and it is probably sound policy when the outsiders do not make a proper effort to gain full control of the content and methods of the biological work in question. Yet, as the history that I examined herein illustrates, all biology contains implicit philosophy, not all of it sound. This suggests that, at least occasionally, it would be desirable to foster interaction across the difficult disciplinary barriers separating biologists and philosophers.

In this chapter, for example, it has become clear that the geographical distinction between nucleus and cytoplasm came to be fraught with a great deal of ideological and philosophical baggage. Together with technical and methodological obstacles impeding a unified approach to heredity and development, divergence in problem choice, and plain old turf battles, the philosophical barriers helped to block communication between geneticists and embryologists for a crucial period lasting some thirty or forty years.

The stereotyped extreme positions that each side ascribed to the other – reflected in the sharp dichotomies set forth previously – made no philosophical and little biological sense. Alternative views, such as I quoted from Driesch and Wilson, were widely available in the literature from the 1890s forward. Perhaps an outsider, properly placed and properly engaged in the debate, might have helped reduce the temptation to erect barriers between genetics and embryology and might have helped to bring some of the principals to recognize that there is more to inheritance than they dreamt of in their philosophies.

## APPENDIX FOR CHAPTER 10: ADDITIONAL AND EXPANDED QUOTATIONS

My doubts as to the validity of Darwin's theory were for a long time not confined to this point alone: the assumption of the existence of *preformed* constituents of all parts of the body seemed to me far too easy a solution of the difficulty, besides entailing an impossibility in the shape of an absolutely inconceivable aggregation of primary constituents. I therefore endeavoured to see if it were not possible to imagine that the germ-plasm, though of complex structure, was not composed of such an immense number of particles, and that its further complication arose subsequently in the course of development. In other words, what I sought was a substance from which the whole organism might arise by *epigenesis*, and not by *evolution*. After repeated attempts in which I more than once imagined myself successful, but all of which broke down when further tested by facts, I finally became convinced that an epigenetic development is an *impossibility*. Moreover, I found an actual *proof of the reality of evolution*, which . . . is so simple that I can scarcely understand how it was possible that it should have escaped my notice so long (Weismann 1893, as quoted in Maienschein 1986).

. . . we reach the following conception. The primary determining cause of development lies in the nucleus, which operates by setting up a continuous series of specific metabolic changes in the cytoplasm. This process begins during ovarian growth, establishing the external form of the egg, its primary polarity, and the distribution of substances within it. The cytoplasmic differentiations thus set up form as it were a framework within which the subsequent operations take place, in a more or less fixed course, and which itself becomes ever more complex as development goes forward . . . (Wilson 1896, as quoted by Sander 1986).

Heredity is to-day the central problem of biology. The problem may be approached from many sides – that of the breeder, the experimenter, the statistician, the physiologist, the embryologist, the cytologist – but the mechanism of heredity can be studied best by the investigation of the germ cells and their development . . .

The comparison of heredity to the transmission of property from parents to children has produced confusion in the scientific as well as in the popular mind . . . [I]n a literal sense parental characteristics are never transmitted to children. Every new individual . . . owes its similarity to its parent to the fact that it was once a part of it, and not to something which has been "transmitted" from one generation to the next. Furthermore from its earliest to its latest stage an individual is one and the same organism; the egg of a frog is a frog in an early stage of development and the characteristics of the adult frog develop out of the egg, but are not transmitted through it by some "bearer of heredity."

Indeed, heredity is not a peculiar or unique principle for it is only similarity of growth and differentiation in successive generations. The fertilized egg cell undergoes a certain form of cleavage and gives rise to cells of particular size and structure, and step by step these are converted into a certain type of blastula, gastrula, larva, and adult. In fact, the whole process of development is one of growth and differentiation, and similarity of these in parents and offspring constitutes hereditary likeness. The causes of heredity are thus reduced to the causes of the successive differentiations of development, and the mechanism of heredity is merely the mechanism of differentiation . . .

Differentiation, and hence heredity, consists in the main in the appearance of unlike substances in protoplasm and their localization in definite regions or cells. Such a definition is as applicable to the latest stages of differentiation, such as the formation of muscle fibers, as it is to the earliest differentiations of the germ cells, and the one is as truly a case of inheritance as is the other . . .

It is known that constructive metabolism, differentiation, and regeneration never occur in the absence of a nucleus. On the other hand, Verworn has shown that the nucleus alone is incapable of performing these functions, and he maintains that the chief role in the life of the cell can not be assigned to either the nucleus or the cytoplasm, but that both are concerned in vital phenomena . . . [In sum], there is good reason to believe that the different substances which appear in the differentiation of a tissue cell arise through the interaction of the nucleus and cytoplasm, and not from either of these alone . . .

Finally, we may conclude that the nucleus plays a less important role in the localization of different substances than in the formation of those substances. Nevertheless, in differentiation, as well as the metabolism, there is every reason to believe that the entire cell is a physiological unit. Neither the nucleus nor the cytoplasm can exist long independently of the other; differentiations are dependent upon the interaction of these two parts of the cell; the entire germ cell, and not merely the nucleus or cytoplasm is transformed in the embryo or larva; and it therefore seems necessary to conclude that both nucleus and cytoplasm are involved in the mechanism of heredity . . .

[In spite of the immense cytological evidence cited in support of the claim that] the chromosomes are the only "bearers of the inheritance material," [the experimental evidence is not decisive] . . .

[On the contrary, equally strong evidence shows] that many fundamental differentiations are found in the cytoplasm of the egg at the time of fertilization and immediately after. As evidences of such differentiations may be cited, (1) polarity and symmetry, (2) differential cleavages, (3) positions and proportions of important organ bases, (4) various types of egg organization, (5) experiments in hybridization . . .

[On the basis of this detailed evidence, it may be concluded that] *at the time of fertilization the hereditary potencies of the two germ cells are not equal, all the early development, including the polarity, symmetry, type of cleavage, and the relative positions and proportions of future organs being predetermined in the cytoplasm of the egg cell, while only the differentiations of later development are influenced by the sperm. In short, the egg cytoplasm fixes the type of development and the sperm and egg nuclei supply only the details* (Conklin 1908, italics in the original).

We have come to look upon the problem of heredity as identical with the problem of development. The word heredity stands for those properties of germ-cells that find their expression in the developing and developed organism. When we speak of the transmission of characters from parent to offspring, we are speaking metaphorically; for we now realize that it is not characters that are transmitted to the child from the body of the parent, but that the parent carries over the material common to both parent and offspring . . .

It may be said in general that the particulate theory [Morgan here includes Mendelism] is the more picturesque or artistic conception of the developmental process. As a theory, it has in the past dealt largely in symbolism and is inclined to make hard and fast distinctions. It seems to better satisfy a class or type of mind that asks for a finalistic solution, even though the solution be purely formal. But the very intellectual security that follows in the train of such theories seems to me less stimulating for further research than does the restlessness of spirit that is associated with the alternative [i.e., physiological, epigenetic, cytoplasm-centered] conception. The purely adventurous character of any explanation offered by the reaction theory seems more in accord with the modern spirit of scientific theory . . .

. . . the facts . . . go far towards showing that the central axis of the chromosome is not lost in the resting nucleus, but remains to become the center of the next chromosome [after a cell division] . . . If we look upon the spinning process of the chromosome as a process by means of which its peripheral substance is thrown out into the nucleus to form the reticulum, and assume that most of it fails to return the next time the chromosome becomes distinct, we have an hypothesis in conformity with many facts at least, and also a view that makes simpler, perhaps, our interpretation of the meaning of the process . . .

If Mendelian characters are due to the presence or absence of a specific chromosome [or part of a chromosome], as Sutton's hypothesis assumes, how can we account for the fact that the tissues and organs of an animal differ from each other when they all contain the same chromosome complex . . . [W]e must be prepared to admit that the evidence is entirely in favor of the view that the differentiation of the body is due to other factors that modify the cells in one way or in another. This consideration is, to my mind, a convincing proof that we have to deal with two sets of factors – the common inheritance of all the cells to produce all the kinds of tissues and organs in the body, and the limitation of that property in the course of development. If the former is due to the chromosomes and the unspecialized parts of the cytoplasm, the latter may be due to the local changes that the relation of the parts to each other calls forth. It might even be argued that since in the development we find no evidence of a sorting out of the chromosomes that produce special parts, the individual chromosomes can not stand each as the representative of these parts, but rather that each part needs the entire set of chromosomes for its normal life (Morgan 1910).

If another branch of zoology that was actively cultivated at the end of the last century had realized its ambitions, it might have been possible to-day to bridge the gap between gene and character, but despite its high-sounding name of *Entwicklungsmechanik* nothing that was really quantitative or mechanistic was forthcoming. Instead, philosophical platitudes were invoked rather than experimentally determined factors. Then, too, experimental embryology ran for a while after false gods that landed it finally in a maze of metaphysical subtleties. It is unfortunate, therefore, that from this source we can not add, to the three contributory lines of research which led to the rise of genetics, a fourth and greatly needed contribution to bridge an unfortunate gap . . .

What has been said so far relates to the action of the gene on the cytoplasm of its own cell – its intracellular action. Those of us working with insects or plants are apt to think of genetic problems in this way, and are inclined to consider mainly the effects that do not reach beyond the cells in which they are produced. But in other groups, especially birds and mammals, the effects of the genes are not always so limited. We are on the threshold of work concerned with the isolation of the so-called sex-hormones, the end-products of the thyroid gland, the pituitary, the thymus, and the substances isolated from the suprarenal bodies (Morgan 1932).

In Wettstein's formulation of the [theory of the] "plasmon" as "bearer of developmental process" . . . we come nearer to the point of view of many biologists who deny that Mendelian genes are the sole representatives of the material of heredity. This denial is generally without experimental foundation; it rests primarily on the need to find a *principle of unity* in organic development. It rests on the disinclination to be forced to treat the organism as an *aggregate* of *determinants*, the complete independence of which cannot guarantee the unity of the organism . . . [A]lthough the various authors supporting this point of view have quite different positions regarding the question whether this substance [i.e., the plasmon] is localized in the nucleus or the cytoplasm, they agree in maintaining that not all developmental stages and phenotypes of organisms are determined by genes, i.e., by discrete entities . . .

Thus the "unitary" conception, according to which the totality of the hereditary material is given in the genes, is opposed by the dualistic conception just described.

Can we refute one or the other of these positions by experiments at this time? Not by means of *classical Mendelism*. For in order to recognize or to localize genes, we need two organisms that differ in some particular traits, and even if we can trace certain fundamental properties (such as chromosome form or size) back to Mendelian genes, it appears to me to be completely impossible to remove the force of the dualists' arguments by analysis of hybrids. Johannsen put it like this: " . . . *By analysis of hybrids we have examined only the clothing*, the underlying organization remains unanalyzed. Whether we will ever be able to strip kinds of organisms [*Gattungen*, a word then used for races, species, and genera in German] of their superficial characteristics in such a way as to reveal the ultimate X of our formulas – a fundamental substance, something quite general and organic, something that, like a homozygote, is not accessible to Mendelian analysis – that remains an unanswerable question (Hertwig 1934, p. 428) [translated by RB, emphases in the original, references omitted].

If as is generally implied in genetic work (although not often explicitly stated), all of the genes are active all the time, and if the characters of the individual are determined by the genes, then why are not all the cells of the body exactly alike?

The same paradox appears when we turn to the development of the egg into an embryo. The egg appears to be an unspecialized cell, destined to undergo a prescribed and known series of changes leading to the differentiation of organs and tissues. At every division of the egg, the chromosomes split lengthwise into exactly equivalent halves. Every cell comes to contain the same kinds of genes. Why, then, is it that some cells become muscle cells, some nerve cells, and others remain reproductive cells? . . .

[The recent results of Spemann on cell interactions and the organizer] are of interest because they bring up once more, in a slightly different form, the problem as to whether the organizer acts first on the protoplasm of the neighboring region with which it comes in contact, and through the protoplasm of the cells on the genes; or whether the influence is more directly on the genes. In either case the problem under discussion remains exactly where it was before. The conception of an organizer has not as yet helped to solve the more fundamental relation between genes and differentiation, although it certainly marks an important step forward in our understanding of embryonic development (Morgan 1935).

The location of genes in the chromosomes, the proof of their linear order, the association of somatic characters with definite points in the chromosomes, in short, the whole development of the gene theory is one of the most spectacular and amazing achievements of biology in our times. The embryologist, however, is concerned more with the larger changes in the whole organism and its primitive systems of organs than with lesser qualities known to be associated with genic action. As Just remarked [in a AAAS symposium, Dec. 30, 1936], he is interested more in the back than in the bristles on the back and more in the eyes than in eye color.

Now that the necessity of relating the data of genetics to embryology is generally recognized and the "Wanderlust" of geneticists is beginning to urge them in our direction, it may not be inappropriate to point out a danger in this threatened invasion.

The prestige of success enjoyed by the gene theory might easily become a hindrance to the understanding of development by directing our attention solely to the genom [sic], whereas cell movements, differentiation and in fact all developmental processes are actually effected by the cytoplasm. Already we have theories that refer the processes of development to genic action and regard the whole performance as no more than the realization of the potencies of the genes. Such theories are altogether too one-sided (Harrison 1937).

Genetic restriction [i.e., differentiation] then depends upon the removal by the nucleus of certain materials from the cytoplasm, leaving others free. The free materials determine the character of the cell . . . With each cleavage each nucleus fixes all material other than that which makes the blastomere what it is . . . Then the potencies for embryo-formation are all present in the uncleaved egg . . .

[The] conception of the action of the genes as unalterably fixed entities can not explain differentiation of development. For how could the genes be responsible for differentiation, if they are the same in every cell [(Lillie 1927)]? Unless the geneticists assume that their genes are omnipotent, we can not understand how the problem of differentiation can be solved by the gene-theory of heredity.

Every cell in an organism becomes what it is because its cytoplasm has freed its particular potencies whilst the nucleus binds all others. These latter would, if left unbound in the cytoplasm, act as obstacles to the display of special potencies (Just 1939, Chapter 11, first and third paragraphs quoted in Gilbert 1988).

[On the basis of the evidence presented in this paper, it is a plausible hypothesis that a developing embryo includes a pre-pattern, perhaps itself controlled or altered by genes, to which certain genes respond by causing (or not causing) the formation of a particular localized differentiated feature such as a bristle.] What was called the *response* of a gene needs to be defined in exact terms. There is some evidence that all genes act, that is participate, in cellular physiological process continuously in every cell. If this evidence is sound, response in differentiation would not signify an awakening from inactivity, but only a rise of activity, or an outcome of activity different from what had occurred. If a necessary prerequisite for a specific differentiation were the production of a specific molecular species, then "response" may mean nothing but linking together or otherwise transforming, by the same enzymatic power which was exerted all through development, molecular groups which had not existed before the pre-pattern provided for their establishment. It is possible to interpret certain well-established differences in the appearance of one and the same narrow band in the giant chromosomes of different tissues of the fly as indicators for differential activity of the same gene in response to different internal environments.

The cellular physiology of differentiation – the specific mechanisms by means of which an embryonic cell is transformed gradually, and at the end usually irreversibly, into a differentiated cell – remains obscure. Some investigators believe that differentiation is based upon alternative stable dynamic equilibria of specific reactions. Others think of differentiation in terms similar to the Roux-Weismann theory of development applied to suspected self-perpetuating particles in the cytoplasm instead of in the nucleus, which become segregated into different parts of an embryo or undergo differential changes in different regions. Either of these views can be linked with the concept which holds that differentiation is the result of response of specific genes to specific pre-patterns.

An excellent basis for this concept has been provided in the studies of Sonneborn and Beale on antigens in the unicellular ciliated animal *Paramecium*. At present three groups of genes are known which determine the antigenic characters of these organisms, but a member of only one group will show its presence at any time by the production of a specific antigen. Which one is expressed depends upon the state of the cytoplasm. There is a cytoplasmic state favorable for the expression of the *g*-genes, another cytoplasmic state favorable for the expression of the *d*-genes, and a third for the *s*-genes. The cytoplasmic state itself is under external control. Three slightly different temperatures may call forth any one of the three states. Imagine a tissue composed of hundreds of *Paramecia*. Apply the three different temperatures to different areas of this tissue. Then in each area one specific gene out of the three concerned would respond to the temperature pre-pattern set up in the tissue, and a patterned differentiation into three antigenically distinguishable types of cells would result (Stern 1954) [references omitted, italics in the original].

## REFERENCES

Allen, G. E. 1974. "Opposition to the Mendelian chromosome theory: The physiological and developmental genetics of Richard Goldschmidt." *Journal of the History of Biology* 7: 49–92.

Allen, G. E. 1986. "T. H. Morgan and the split between embryology and genetics, 1910–1935." In *A History of Embryology*. eds. T. J. Horder, J. A. Witkowski, and C. C. Wylie. Cambridge: Cambridge University Press, 113–46.

Benson, K. R., J. Maienschein, and R. Rainger, eds. 1991. *The Expansion of American Biology*. New Brunswick, NJ: Rutgers University Press.

Boveri, T. 1887. "Die Bildung der Richtungskörper bei *Ascaris megalocephala* und *Ascaris lumbricoides*." *Jenaische Zeitschrift für Naturwissenschaft* 21: 423–515.

Boveri, T. 1892. "Über die Entstehung des Gegensatzes zwischen den Geschlechtszellen und die somatischen Zellen bei *Ascaris megalocephela*." *Sitzungsberichte der Gesellschaft für Morphologie und Physiologie* 8: 114–25.

Boveri, T. 1902. "Die Befruchtung und Teilung des Eies von *Ascaris megalocephala*." *Jenaische Zeitschrift für Naturwissenschaft* 22: 685–882.

Boveri, T. 1910. "Über die Teilung centrifugierter Eier von *Ascaris megalocephala*." *Wilhelm Roux Archiv für Entwicklungsmechanik der Organismen* XXX: 101–25.

Burian, R. M., and J. Gayon. 1999. "The French school of genetics: From physiological and population genetics to regulatory molecular genetics." *Annual Review of Genetics* 33: 313–49.

Burian, R. M., J. Gayon, and D. T. Zallen. 1988. "The singular fate of genetics in the history of French biology, 1900–1940." *Journal of the History of Biology* 21: 357–402.

Burian, R. M., J. Gayon, and D. T. Zallen. 1991. "Boris Ephrussi and the synthesis of genetics and embryology." In *A Conceptual History of Embryology*, ed. S. Gilbert. New York: Plenum, 207–27.

Conklin, E. G. 1908. "The mechanism of heredity." *Science* 27: 89–99.

Darlington, C. D. 1939. *The Evolution of Genetic Systems*. Cambridge: Cambridge University Press.

Darlington, C. D. 1944. "Heredity, development, and infection." *Nature* 154: 164–9.

Driesch, H. 1894. *Analytische Theorie der organischen Entwicklung*. Leipzig: Wilhelm Engelmann.

Ephrussi, B. 1958. "The cytoplasm and somatic cell variation." *Journal of Cellular and Comparative Physiology* 52, suppl. 1: 35–53.

Gayon, J., and R. M. Burian. 2000. "France in the era of Mendelism (1900–1930)." *Comptes rendus de l'Académie des Sciences, Paris. Série III, Sciences de la Vie/Life Sciences* 323, no. 12: 1097–107.

Gilbert, S. F. 1988. "Cellular politics: Ernest Everett Just, Richard B. Goldschmidt, and the attempt to reconcile embryology and genetics." In *The American Development of Biology*, eds. R. Rainger, K. R. Benson, and J. Maienschein. Philadelphia: University of Pennsylvania Press, 311–46.

Gilbert, S. F. 1991. "Induction and the origins of developmental genetics." In *A Conceptual History of Modern Embryology*, ed. S. F. Gilbert. New York: Plenum, 181–206.

Hamburger, V. 1980. "Embryology and the modern synthesis in evolutionary theory." In *The Evolutionary Synthesis*, eds. E. Mayr and W. Provine. Cambridge: Harvard University Press, 97–111.

Harrison, R. G. 1937. "Embryology and its relations." *Science* 85: 369–74.

Harrison, R. G. 1940. "Cellular differentiation and internal environment." In *The Cell and Protoplasm*, ed. F. G. Moulton. Washington, DC: The Science Press, 77–97.

Hertwig, P. 1934. "Probleme der heutigen Vererbungslehre." *Die Naturwissenschaften* 25: 425–30.

Just, E. E. 1939. *The Biology of the Cell Surface*. Philadelphia: P. Blakiston's Son and Co.

Lillie, F. R. 1927. "The gene and the ontogenetic process." *Science* 64: 361–8.

Maienschein, J. 1981. "Shifting assumptions in American biology: Embryology, 1890–1910." *Journal of the History of Biology* 14: 89–113.

Maienschein, J. 1983. "Experimental biology in transition: Harrison's embryology, 1895–1910." *Studies in History of Biology* 6: 107–27.

Maienschein, J. 1986. "Preformation or new formation – or neither or both?" In *A History of Embryology*, eds. T. J. Horder, J. A. Witkowski, and C. C. Wylie. Cambridge: Cambridge University Press, 73–108.

Maienschein, J. 1987. "Heredity/development in the United States, circa 1900." *History and Philosophy of the Life Sciences* 9: 79–93.

Maienschein, J. 1990. "Cell theory and development." In *A Companion to the History of Modern Science*, eds. R. C. Olby, G. N. Cantor, J. R. R. Christie, and M. J. S. Hodge. London: Routledge, 357–73.

Maienschein, J. 1991. *Transforming Traditions in American Biology, 1880–1915*. Baltimore and London: Johns Hopkins University Press.

Morgan, T. H. 1910. "Chromosomes and heredity." *American Naturalist* 44: 449–96.

Morgan, T. H. 1917. "The theory of the gene." *American Naturalist* 51: 513–44.

Morgan, T. H. 1926. *The Theory of the Gene*. New Haven: Yale University Press.

Morgan, T. H. 1932. "The rise of genetics." *Science* 76: 261–7, 285–9.

Morgan, T. H. 1934. *Embryology and Genetics*. New York: Columbia University Press.

Morgan, T. H. 1935. "The relation of genetics to physiology and medicine (Nobel Lecture, Stockholm, June 4, 1934)." *Scientific Monthly* 41: 5–18.
Morgan, T. H., A. H. Sturtevant, H. J. Muller, and C. B. Bridges. 1915. *The Mechanism of Mendelian Heredity*. New York: Henry Holt and Co.
Nanney, D. L. 1958. "Epigenetic control systems." *Proceedings of the National Academy of Sciences, USA* 44: 712–17.
Olby, R. C. 1990. "The emergence of genetics." In *A Companion to the History of Science*, eds. R. C. Olby et al. London: Routledge, 521–36.
Rainger, R., K. R. Benson, and J. Maienschein, eds. 1988. *The American Development of Biology*. Philadelphia: University of Pennsylvania Press.
Roux, W. 1883. "Über die Bedeutung der Kerntheilungsfiguren: Eine hypothetische Erörterung." Leipzig: Engelmann.
Roux, W. 1991. "On the significance of the figures of nuclear division: A hypothetical exposition, translated and introduced by Richard M. Burian and Marjorie Grene." *Evolutionary Biology* 25: 427–44.
Sander, K. 1986. "The role of genes in ontogenesis – evolving concepts from 1883–1983 as perceived by an insect embryologist." In *A History of Embryology*, eds. T. J. Horder, J. A. Witkowski, and C. C. Wylie. Cambridge: Cambridge University Press, 363–95.
Sapp, J. 1983. "The struggle for authority in the field of heredity, 1900–1932: New perspectives on the rise of genetics." *Journal of the History of Biology* 16: 311–42.
Sapp, J. 1986. "Inside the cell: Genetic methodology and the case of the cytoplasm." In *The Politics and Rhetoric of Scientific Method*, eds. J. A. Schuster and R. R. Yeo. Dordrecht, Holland: D. Reidel, 167–202.
Sapp, J. 1987. *Beyond the Gene: Cytoplasmic Inheritance and the Struggle for Authority in Genetics*. New York: Oxford University Press.
Stern, C. 1954. "Two or three bristles." *American Scientist* 42: 213–47.
van Beneden, E., and A. Neyt. 1887. "Nouvelles recherches sur la fécondation et la division mitosique chez l'Ascaride mégalocéphale." *Bulletin de l'Académie Royale des Sciences, des Lettres et des Beaux-Arts de Belgique* 14: 215–95, Pl. I–VI.
Weismann, A. 1892. *Das Keimplasma. Eine Theorie der Vererbung*. Jena: Fischer.
Weismann, A. 1893. *The Germ-Plasm: A Theory of Heredity*. New York: The Macmillan Co.
Wilson, E. B. 1896. *The Cell in Development and Inheritance*. New York: The Macmillan Co.
Wollman, E. 1928. "Bactériophagie et processus similaires: Hérédité ou infection?" *Bulletin de l'Institut Pasteur* 26: 1–14.
Wollman, E., and E. Wollman. 1925. "Sur la transmission 'parahéréditaire' de caractères chez les bactéries." *Comptes Rendus de la Société de Biologie* 93: 1568–9.

# 11

# On Conflicts between Genetic and Developmental Viewpoints – and Their Attempted Resolution in Molecular Biology[1]

> Embryologists and geneticists never used to see eye to eye. As I have indicated in this book the two disciplines have now become united in a new subject formed by the fusion of developmental genetics with molecular biology (Lawrence 1992, p. 195).

This chapter concerns the apparent rapprochement, now underway, between genetics and developmental biology. For there to be a rapprochement, of course, there must have been long-standing disagreements. The disagreements between embryology and genetics bear all the stigmata of a deep discordance between research traditions built on conflicting assumptions and practices. They have been discussed by many historians of science and shown to rest on conceptual and institutional differences, on differences in the experimental and field practices of geneticists and embryologists, and on the distinctive behaviors of the biological materials they traditionally employed.[2] Suffice it to say that the disagreements were based in part on the absolute inability of geneticists to show how genes could account for the *Bauplan* of an organism

[1] Originally presented at the 10th International Congress of Logic, Methodology, and Philosophy of Science in Florence, August 1995, and published as "On Conflicts Between Genetic and Developmental Viewpoints – and Their Resolution in Molecular Biology" in *Structure and Norms in Science*, vol. 2 of the Congress proceedings edited by M. L. Dalla Chiara, K. Doets, D. Mundici, and J. van Bentham (Dordrecht, Holland: Kluwer, 1997), pp. 243–64. Reprinted with kind permission of Kluwer Academic Publishers. In preparing this chapter, I benefited greatly from the penultimate draft of Gilbert, Opitz, and Raff (1996) and from discussions with Prof. Gilbert. I am grateful to him for sharing this work with me, for permission to utilize it herein, and for permission to reproduce Figures 3 and 7. Thanks also to Dr. Opitz and participants at the ICLMPS session in Florence for constructive comments, to Prof. Walter Gehring for supplying the electron micrographs reproduced in the original article in Figures 5 and 6, and to the American Association for the Advancement of Science for the versions used on this occasion.

[2] See, for example, Allen (1986), Harwood (1984, 1993), Keller (1995, Chapter 1), Maienschein (1986), Sander (1986), Sapp (1983, 1986, 1987), and Chapters 2 and 10.

and on their failure to give any weight to such phenomena as cytoplasmic gradients in the egg, polarities in the egg and the early embryo, cell death in organogenesis, and so on.[3] These complaints against genetics were just and remain so, although that fact does not reveal whether they are biologically or philosophically important.

The rapprochement of interest here results from the application of powerful new molecular techniques to basic biological problems and from fundamental conceptual revisions now under way in both genetics and developmental biology. The conceptual changes, which are strongly influenced by recent findings employing molecular techniques, depend in part on paying increased attention to the many ways in which genes are activated. Briefly, one tendency that deserves our attention treats development in terms of the activation of genes by complex higher level systems rather than treating it as a passive consequence of gene action. I describe a few illustrative molecular findings, discuss conflicting interpretations of the putative reconciliation, highlight some conceptual issues, and begin an examination of some ways in which molecular findings require changes in basic concepts and patterns of explanation in both genetics and developmental biology. The discussion leads me to suggest that the hoped-for reconciliation between genetics and developmental biology will likely undermine the prior conceptual foundations of each discipline.

In particular, I suspect that the reconciliation between developmental biology and genetics will not rest on a theoretical unification of the prior tenets of these disciplines regarding the control of organismal development. Rather, the reconciliation will require partial abandonment of a number of key commitments on both sides. Such as it is, the ongoing reconciliation rests on a detailed understanding of the bits and pieces of machinery out of which organisms are built and how they work. For the present, at least, we understand development better by working from the molecular bottom up rather than from the organismic (or some theoretical) top down. But, contentiously, the molecular bottom is not properly construed either in terms of the master genes for which many geneticists hoped or in terms of the gene-independent gradients, pre-patterns, and morphogenetic fields for which many embryologists hoped. Rather, the molecular bottom involves bounded, sometimes very short-lived interactive

[3] Cf. Sander (1995) for a brief account of some little-known early work along these lines and Morgan (1934, 1935) for one of the first widely known articulations of an in-principle means of combining cytoplasmic gradients with rigid nuclear and genetic determination of all potentialities of the organism. I do not know any good histories of such concepts as that of a morphogenetic field. Some useful materials are included in Horder, Witkowski, and Wylie (1986), especially Wolpert (1986); see also de Robertis (1994), de Robertis, Mortita, and Cho (1991), Gilbert (1991), and Opitz (1985).

regulatory networks built by extraordinarily close interactions between genes and proteins. These networks, in turn, have extraordinarily complex and diverse relations to the many diverse pieces of molecular machinery that produce the phenotypic structures whose development the networks regulate. Many of the regulatory networks operate within specific geographies, most of which are not yet laid down in the earliest embryos; some of them are open to "resetting" only during brief critical periods. The control systems both require and build polarities and gradients that define the boundaries within which further control systems operate. Thus, geographically and temporally bounded domains, gradients, and control systems all play an inescapable role in ontogeny. Accordingly, aspects of both traditional embryology and traditional genetics are represented in the detailed understanding of developmental processes. What remains obscure and is hotly contested is how these aspects can be put together coherently.

We are far from understanding the presumptive global controls by means of which the bodies of organisms are integrated during the course of development. We do not understand how the many interlocking processes are coordinated so as to allow the various structures to fit together and achieve functional coordination among body parts. But, it is already clear that many highly modular local control mechanisms go into the making of an integrated organism. The resulting picture is, thus, vastly different in detail and in character than was anticipated by classical genetics or by classical embryology – or, for that matter, by molecular genetics before the 1980s. At the same time, we know, for example, that strongly conserved regulatory devices contribute to the control of such processes as the making of body segments and eyes. In the specific case of eyes, discussed in a later section, the same control system regulates the beginning steps for making such radically different eyes as the compound eye of a fruitfly and the eye of a mouse or a human being. This example, like many others, raises and exemplifies enormous conceptual problems, among them the interplay of the universal and the specific in control of development. Indeed, it serves as a useful vehicle for posing a number of the conceptual problems raised by recent molecular work on development. To illustrate these problems, I discuss the difficulty in making sense of the distinction between homology and analogy when, as I just hinted is the case for eyes, homologous processes controlled by homologous genes serve to regulate the construction of such seemingly nonhomologous but analogous structures as dipteran and mammalian eyes.

I maintain that the integration of organisms is not readily understood within a unified account of the sorts that many geneticists and developmental biologists have hoped to forge. Organisms are, in many respects, like complex machines built from a diversity of different, hierarchically interacting, partly independent multilevel units or modules on different scales. The boundaries of important units in development only sometimes correspond with our prior intuitions and with obvious phenotypic markers. The development of many modules is under partly autonomous control, modulated by extremely diverse interactions with neighboring units during ontogeny. I put a bit of biological flesh on this skeletal picture in an attempt to set forth an account of some of the obstacles to forging a unified and satisfactory theory of development on any familiar theoretical foundation. But, let us turn to a description of some recent findings in molecular genetics and get to work.

## SOME RECENT FINDINGS IN MOLECULAR DEVELOPMENTAL GENETICS

It would be easy to get lost in the maze of details of recent molecular work bearing on the genetics of development. Any attempt to summarize this work must be incredibly unrepresentative of a large, complex, and flourishing enterprise, one that employs extremely powerful tools to examine the details of extremely complicated and diverse systems. Nonetheless, the following sketch is meant to convey the spirit of one main line of findings. It discusses some work that is likely to require fundamental revisions of concepts in developmental biology, evolutionary biology, and genetics. The study of work like that discussed in this chapter may help us to rethink – and perhaps revise – the patterns of explanation we employ for many biological phenomena.

Major surprises, with wholly unexpected ramifications, stem from the discovery, sequencing, and functional analysis of such nucleic acid sequences as those of so-called homeoboxes, first characterized in 1982.[4] Homeoboxes are sequences, typically of 180 nucleotides (often plus introns), encoding a sequence of 60 amino acids. These form a so-called homeodomain in the corresponding protein. Homeodomains are transcription factors that regulate interactions between the proteins that contain them and DNA; they ensure that the proteins occupy specific sites on DNA. By occupying the specific sites,

[4] For brief accounts of the discovery of homeoboxes in late 1982, see Lawrence (1992 #15276) and Gehring (1994). All good recent textbooks of molecular genetics, developmental biology, or evolution already contain an account of homeoboxes, but detailed knowledge is moving very fast in the primary literature. Gehring has since published a useful book (Gehring 1998).

the proteins in question alter the likelihood that nearby stretches of DNA will be transcribed.[5]

Homeoboxes have an enormously broad phylogenetic distribution; they are found in all metazoans and are known in a number of fungi and plants. So far as is known, they occur exclusively in genes regulating cell fate or producing proteins that mark segment boundaries during development; these often produce patterns on which body parts are laid down. The proteins they encode play numerous roles in processes regulating the expression of concerted groups of genes and/or the formation of structural units in the bodies of animals. The amino-acid sequences in homeodomains are enormously conserved in evolution – for example, fifty-nine of the sixty amino acids in a particular homeobox of the toad *Xenopus* are the same as those in the homeobox of the *Antennapedia* gene in *Drosophila* (Lewin 1990, p. 770).[6]

Homeoboxes were originally discovered in *Drosophila*; their name stems from the fact that the original homeoboxes were found in genes in which homeotic mutations occur. "Homeotic mutations" (a term coined by Bateson) are those that replace one body part with another. One of the most famous of these, the bithorax mutation first discovered by Calvin Bridges and T. H. Morgan in 1915 and worked on extensively since 1948, primarily by Ed Lewis, replaces the balancers (halteres) of *Drosophila* with wings, yielding a

[5] Sixty amino acids are the canonical number in homeodomains. A few shorter homeodomains are known and there are both upstream and downstream extensions in some cases (Bürglin 1994). The homeodomain "was the first of the DNA binding motifs to be identified in *Drosophila* . . . The sequence encodes four main $\alpha$-helices linked by turns. The third of these helices fits into the major groove of the DNA, 'the recognition helix,' and several amino acids on one face of this helix contact the bases of the DNA. The structure has been studied and this has helped identify those amino acids that make particular contacts, not all of which lie inside the third helix. When sequences from different genes are compared, these contacting amino acids are rather variable and may well determine which binding sites the protein seeks out. If one of these amino acids is altered, so are the *in vitro* binding properties of the domain" (Lawrence 1992, p. 110). Current results suggest that there are generally three helices, although the third often has a kink, thus producing what was and sometimes is counted as a fourth helix (see Bürglin 1994, p. 28).

[6] The pairwise sequence similarity of twenty-one homeodomains from a total of 346 homeodomain amino-acid sequences analyzed in Bürglin (1994) ranges from 20 to 67 percent. The twenty-one homeodomains were chosen to represent distinct classes; thus, they differ far more than randomly chosen homeodomains. The twenty-one domains chosen come from *Drosophila*, mouse, a brachiopod, *caenorhabditis* (a nematode), a yeast, and *arabidopsis* (a plant). Among all 346 homeodomains, seven of the sixty sites in the homeodomain are occupied by only one or two amino acids, six have three to five amino acids, and thirteen have six to nine amino acids in common. The substitutions in these sites are also conservative in that they are typically function-preserving (e.g., hydrophobic amino acids are replaced by hydrophobic amino acids). Thus, twenty-six sites are especially conservative and can serve as diagnostic tools for recognizing homeodomains. Parallel claims apply for the corresponding nucleotides of the DNA, although the nucleotide substitution rate is, of course, higher than that of amino acids.

four-winged fly. (This mutant is discussed at greater length in Chapter 12.) By now, more than one hundred *Drosophila* genes are known to contain homeoboxes. Every one of them appears to play some role in regulating segment boundaries or cell or segment fate. Proteins containing homeodomains are often part of extraordinarily intricate networks of genes and proteins that regulate the fate of a phenotypically identifiable unit. At the same time, it is important to know that the boundaries defined by the proteins do not necessarily correspond to the familiar phenotypic units and that alterations in one protein in a given cascade can drastically alter the boundaries laid down by another (Figure 1).[7] The best understood example, far richer in detail than can be hinted at here, illustrates this point. It concerns the genes in the bithorax complex of *Drosophila*, which form part of the apparatus that regulate the division of the embryo into segments and determines the fates of those segments.

Homeotic mutants provide important clues to the segmentation of a body. In *Drosophila* and other segmented organisms, it is possible for the phenotype of one segment to be replaced with that of another segment – witness the bithorax mutant fly (*di*pteran) with four wings (*quadri*pteran?) or mutations in mice that result in cervical vertebrae occupying a place that should be occupied by thoracic vertebrae.[8] But, there are also severe restrictions on such replacements – for example, in a large class of cases, posterior parts can be replaced by anterior parts but not vice versa. In the case of bithorax, there is an exact correspondence between the order in which certain contiguous genes are turned on and the determination of the body segments proceeding from the anterior end of the embryo toward its posterior end. I can only give a primitive first approximation account of this story, worked out largely thanks to the ingenious persistence of Lewis, recently followed by many others.

The three genes in the bithorax series are located contiguously on the third chromosome. These are turned on sequentially in both time and space along an anterior-to-posterior axis in the embryo. Thanks to a series of alternative splices, a series of ten distinct products are made in this way (Figure 2). This process depends heavily on protein–DNA interactions. Formation of the normal sequence of body segments requires that the homeoboxes of the

[7] For example, the genes that define the boundaries between what are now called parasegments in the early *Drosophila* embryo are shifted one-half segment away from the segments of the adult fly. This became clear only when it became possible to visualize the boundaries in question by making images showing the distribution of the proteins produced by the homeobox-containing genes, such as *engrailed, fushi tarazu*, and *paired*, which demarcate parasegment boundaries starting at about the third hour after the egg is laid.

[8] John Opitz (personal communication, 1995) reported that it had not yet been demonstrated that such mutations, which are quite rare in humans, are associated with genes containing homeoboxes.

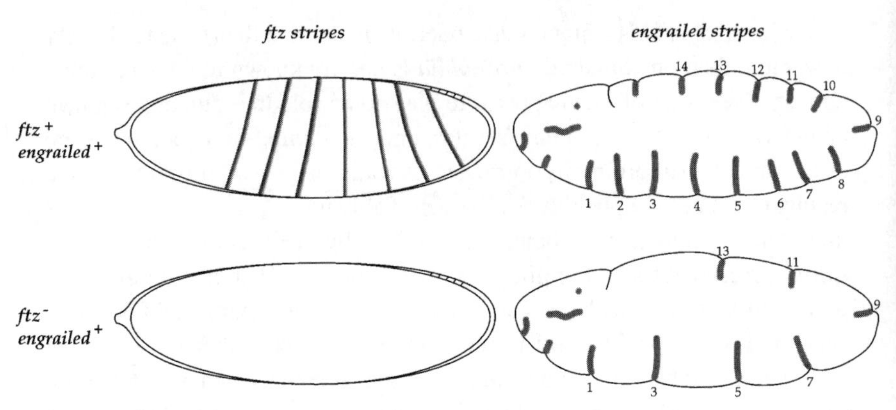


Fig. 1. Striped spatial patterns of the products of the *fushi tarazu* (left) and *engrailed* (right) genes about 3–4 hours after a *Drosophila* egg is laid. Top row, both genes wild type. Bottom row, no *fushi tarazu* gene product. The pattern of engrailed gene product stripes is entirely changed in the absence of the *fushi tarazu* gene product. The half-life of the *fushi tarazu* product is about 5 minutes, so the timing of the interaction between the two gene products is critical for establishing the correct boundaries for further events. This timing dependence is typical of interactions affecting segment (and other geographical) boundaries within the embryo. Reproduced, by kind permission of Blackwell Scientific Publishers, from Figure 4.7 of Peter A. Lawrence, *The Making of a Fly: The Genetics of Animal Design*. Oxford: Blackwell Scientific Publications, 1992.

bithorax genes each specify the exact sequences of amino acids required for the protein–DNA interactions that occur in normal development. When something goes wrong with this process, one does not obtain the normal sequence of thoracic segments. Equally interesting, part of the apparatus that regulates whether these genes are turned on is a feedback loop requiring that a protein boundary be established at the midpoint of the preceding segment – or, rather, of the preceding parasegment – in the series.[9] The product of the preceding regulatory gene in the series is one of the proteins required to be present at that boundary in order to turn on the next gene in the series. In

[9] Thus, phenotypic segments in arthropods are divided into anterior and posterior half-segments. It is with respect to these boundaries that the *para*segments are shifted one-half segment anterior to phenotypic segments. Parasegments appear to be the natural units for determining segment fate in the early embryo (see, e.g., Gilbert 1988, p. 638 ff.; Lawrence 1992, p. 91 ff.; or virtually any recent standard text). Parasegments are laid down first; the boundaries within them, which define segment boundaries, are laid down in a subsequent step. The details of the bithorax system with respect to boundary formation and parasegment fate determination are important but cannot be adequately summarized here. A useful summary is found in Lawrence (1992, p. 111 ff.). In fact, the three gene products need to be produced in the correct spatial sequence to correctly posteriorize *para*segments, not segments; interactive effects of the gene products characterize and determine some parasegments rather than simply the presence of one, two, or three of the gene products.

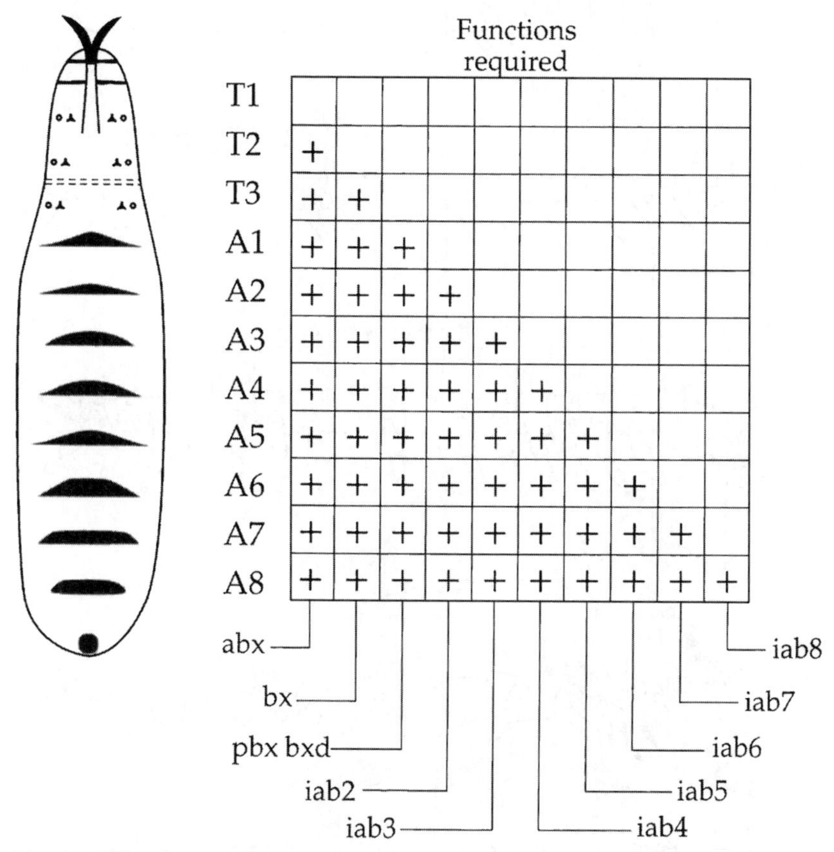


Fig. 2. Ed Lewis's proposal, largely confirmed, for the manner in which the genes of the bithorax complex specify distinct fates for the thoracic and abdominal segments of *Drosophila* and other insects. Successive genes are activated sequentially both in space and in time. The function of an additional gene product in the series is required to specify the identity of each succeeding segment. Redrawn after Fig. 38.15 of B. Lewin, *Genes IV*. Oxford: Oxford University Press, 1990.

consequence, in the absence of the appropriate anterior function, it is often not possible to produce either the anterior *or the adjacent posterior* part and it is never possible to place the posterior part in front of the anterior one. For this reason, although, balancers (halteres) – a more posterior structure – have been replaced by wings in the bithorax mutation, no case is known of halteres, the posterior part, being placed in front of wings.[10]

[10] The production of wings can be blocked or preempted in various ways. Accordingly, it is possible to have halteres in the absence of wings but not in the absence of the expression of the genes required for determination of a segment as a wing-bearing segment.

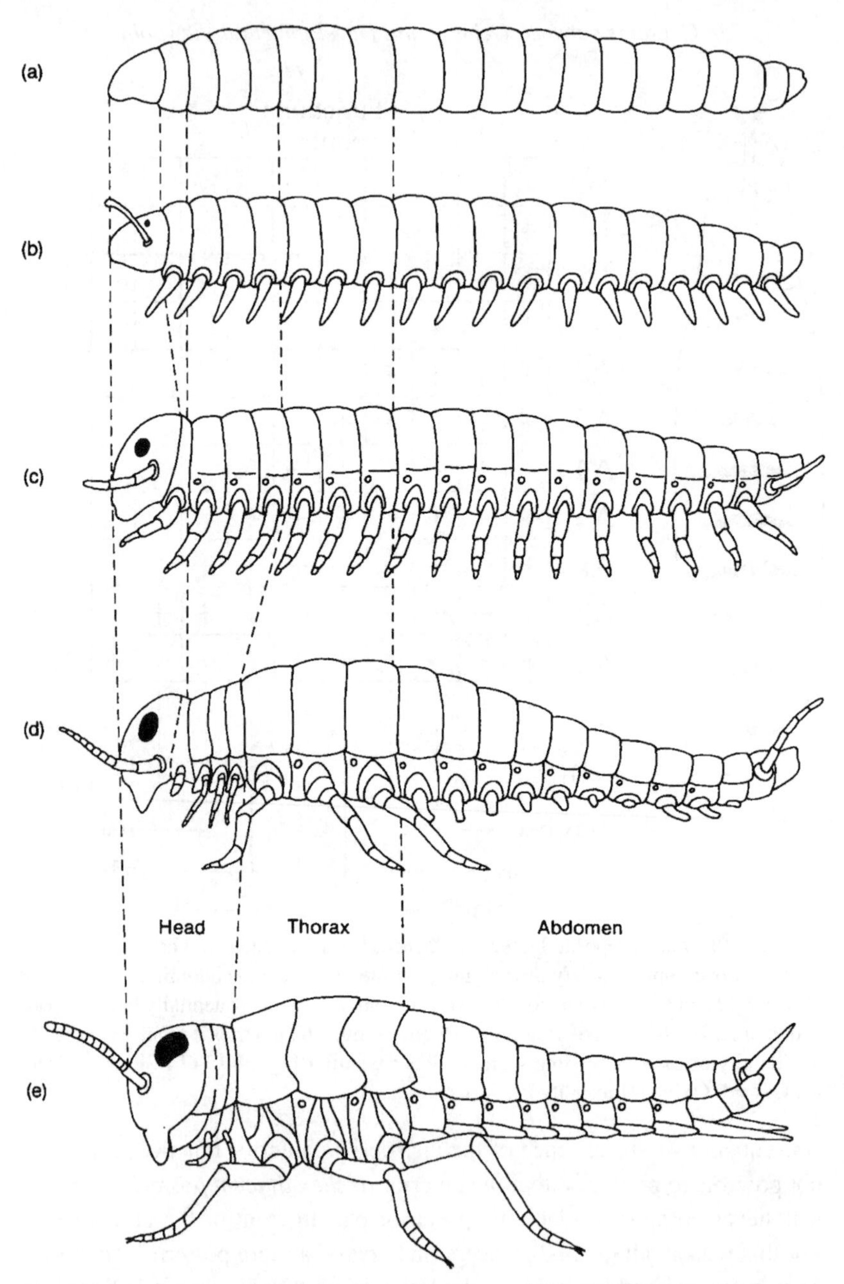


Fig. 3. Diagram of evolution of segmental patterns of insects from some of their ancestors, as hypothesized by Raff and Kaufman 1983, based on the addition of homeotic gene functions. The animals depicted are (a) an annelid, (b) an onycophoran, (c) a myriapod, (d) an apterygote, and (e) a pterygote. Redrawn after Fig. 8–14 of R. A. Raff and T. C. Kaufman, *Embryos, Genes, and Evolution: The Developmental–Genetic Basis of Evolutionary Change*. New York: Macmillan, 1983.

In general, before a homeotic mutation can have its characteristic effect, there must be defined boundaries that determine the geographical limits of its action. But there are many systems that yield homeotic changes and many relevant boundaries, determined by additional control systems. For example, the specification of whether a part of a segment is proximal or distal is independent of the determination of whether that segment yields a leg, an antenna, or a wing. Thus, in homeotic mutants in which parts of an antenna are homeotically transformed into parts of a leg, the transformed proximal parts of the antenna are always proximal parts of a leg and the transformed distal parts of the antenna are always distal parts of the leg.

A further finding sharpens the connection with issues about homology. There is now strong molecular evidence, nearly universally accepted, that the three bithorax genes, together with five "antennapedia" genes, were initially produced by gene duplication followed by differentiation of function. (The same is true of many other groups of genes involved in determining segments.) That is, these two groups of contiguous genes match the classic nineteenth-century definition of Owen for being serial homologues; an initial gene reproduced first by "vegetative repetition" (i.e., gene duplication), followed by subsequent differentiation (Gilbert 1980). The genes of the control system of which we have been talking manifest serial homology at the genetic level. Their relationship is reminiscent of that of the vertebrae in Owen's diagram of the archetypical vertebrate; like the differentiated vertebrae of the various vertebrates, the current genes result from differentiation of the iterated ancestral gene. And, in precise parallel with this, it is likely that an annelid ancestor yielded insects by the specification of different functions per segment (Figure 3). The different segment identities depend on the activation of serially homologous genes that, themselves, have evolved by vegetative repetition followed by differentiation while in partial, though indirect, interaction with one another.[11]

There are yet more interesting surprises about the segment-determining genes that, taken together, imply that these genes are also phylogenetically homologous (i.e., exhibit what Owen called special homology) to genes that carry out similar segmentation functions in nearly all metazoans, from flatworms to mammals.[12] Because of space limitations, I cannot go into great detail. But, examination of Figure 4 helps to summarize some salient points.

[11] This topic is discussed further in the next chapter.

[12] Early in the cloning of homeoboxes, it was found that vertebrates have homeoboxes. In the late 1980s, it became clear that these occur in genes that are molecularly similar to the genes that control early embryonic development in *Drosophila*, including those of the bithorax complex. It also become clear that certain genes and gene clusters controlling segmentation in *Drosophila*

| Anterior → Posterior | | | | | | | | | |
|---|---|---|---|---|---|---|---|---|---|
| Fly Ant-C | lab | pb | Dfd | Scr | Antp | | | | |
| Fly Bx-C | | | | | | Ubx | AbdA | AbdB | |
| Mouse Hox-2 | 2.8 | 2.7 | 2.6 | 2.1 | 2.2 | 2.3 | 2.4 | 2.5 | |
| Mouse Hox-1 | 1.6 | 1.5 | 1.4 | 1.3 | 1.2 | 1.1 | | 1.7 | |
| Mouse Hox-3 | | | | 6.2 | 6.1 | | 3.1 | 3.2 | 3.3 |
| Mouse Hox-5 | | 4.1 | 5.1 | | | | | 5.2 | 5.3 |

Fig. 4. Correspondence between the *Drosophila antennapedia* and *bithorax* gene complexes and the clusters of homeotic Hox genes on four different mouse chromosomes. In both organisms, the genes are expressed sequentially, going from left (the 3′ end of the DNA molecule) to right (5′ end). This is also the order in which the genes are expressed in the anterior to posterior axis of the organism. The genes of the Hox clusters are aligned with the fly genes according to the sequence similarity of the homeodomains of the genes in the respective clusters. Redrawn after Fig 38.19 of B. Lewin, *Genes IV*. Oxford: Oxford University Press, 1990. For a more recent version of this figure, see de Robertis (1994, p. 14), which provides the new designations of the Hox genes and additional information regarding parallels between the genes and their control systems. See also Fig. 12–7.

There is a precise sequential correspondence between four series of genes containing homeoboxes in mice and the antennapedia and bithorax series in *Drosophila*. If one were to line up the genes in each group by examining only the degree of sequence similarity in their homeoboxes, one would obtain exactly the same linear ordering that, in fact, obtains. Furthermore, just as in *Drosophila*, the mouse genes are expressed in the mouse embryo sequentially, with the leftmost (3′) genes in the figure being expressed anteriorly; as one proceeds to the right, each gene is expressed more posteriorly. Yet, further, the genes in mice are "turned on" sequentially in time and specify segment identity in roughly the way that they do in *Drosophila*.[13] Yet, further, the enhancers of the mouse genes serve as enhancers of the *Drosophila* genes and vice versa. (For examples, see Malicki et al. 1992; McGinnis, Kuziora, and McGinnis 1990.) These genes exhibit homology of function and arrangement; the sequence similarities we have mentioned are not accidental but rest on a deep homology of some as-yet ill-understood sort. (For further details and references, see Gehring 1994; Gilbert, Opitz, and Raff 1996; Lawrence 1992.) The recent evidence, reviewed by Gilbert, Opitz, and Raff (1996), suggests that this sequential system is virtually universal in metazoa. Assuming that all this

are contiguous on a single chromosome and that their molecular mates in mice and other animals occur in the same serial order as in *Drosophila*.

[13] This claim is currently disputed; the timing of events in vertebrates that determine sequence identity is apparently different in vertebrates than in *Drosophila*.

stands up to further scrutiny, they summarize the results dramatically with the somewhat simplified formula: "Every animal has similar genes, has them in the same chromosomal order, and uses them to specify the same relative positions along the anterior-posterior axis" (Gilbert, Opitz, and Raff 1996, p. 364). The underlying idea here goes back not to Cuvier but to Geoffroy St.-Hilaire (Arendt and Nübler-Jung 1994; Panchen 2001); the specification of the sequence of segments fundamental to the structure of animal, the basis for specifying their architecture or *Bauplan*, is not different among the four *embranchements* – it is the same in *all* animals.

A final example will help make the conceptual issues stand out more sharply both because it is so dramatic and because it breaks one of the general rules about homeotic genes. The example, based on work done in Gehring's laboratory in Basel, concerns the machinery for making eyes. One of the most exciting new findings (Halder, Callaerts, and Gehring 1995) concerns three genes, clearly homologous, denominated *ey* (eyeless) in *Drosophila*, *Sey* (small eye) in mice, and *Aniridia* (no iris) in humans.[14] In each of these animals, the homologous gene appears to serve as a switch to turn on the entire cascade of phenomena required for formation of eyes. Homologues are also present in ribbon worms, ascidians, cephalopods – probably in nearly all metazoa. The sequence similarities of homeodomains of the mouse and *Drosophila* proteins is 90 percent and that of an associated transcription factor affecting protein–DNA interactions and DNA transcription (called a paired domain) is 94 percent.[15] Remarkably, genetically engineered targeted ectopic expression of *ey* in *Drosophila* imaginal disks induces the formation of eyes independent of locational information – that is, independent of whether the gene is expressed in antenna, leg, wing, or haltere-forming tissue (Figure 5). This breaks the rule, suggested previously, that homeotic mutants require additional positional information to yield one body part rather than another; in this respect, *ey* behaves differently than the other homeotic genes about which I know. (On the other hand, the eye is anterior to any of the other organs in question; perhaps something in the cascade initiated by the eye-forming switch simply blocks or reverses the action of posteriorizing gene products.)

[14] The terminology has since shifted. For reasons that will become clear in the text, these genes are considered version of the same gene. It is one of a family of genes (see the next chapter) and is denominated *Pax6*.

[15] The paired domain, incidentally, is a member of a family of transcription factors distinct from the homeodomain; the combination of both domains on a single protein "illustrate[s] evolutionary tinkering at the molecular level" (Quiring et al. 1994, p. 785) – here tinkering with modular control systems to yield a novel control system.

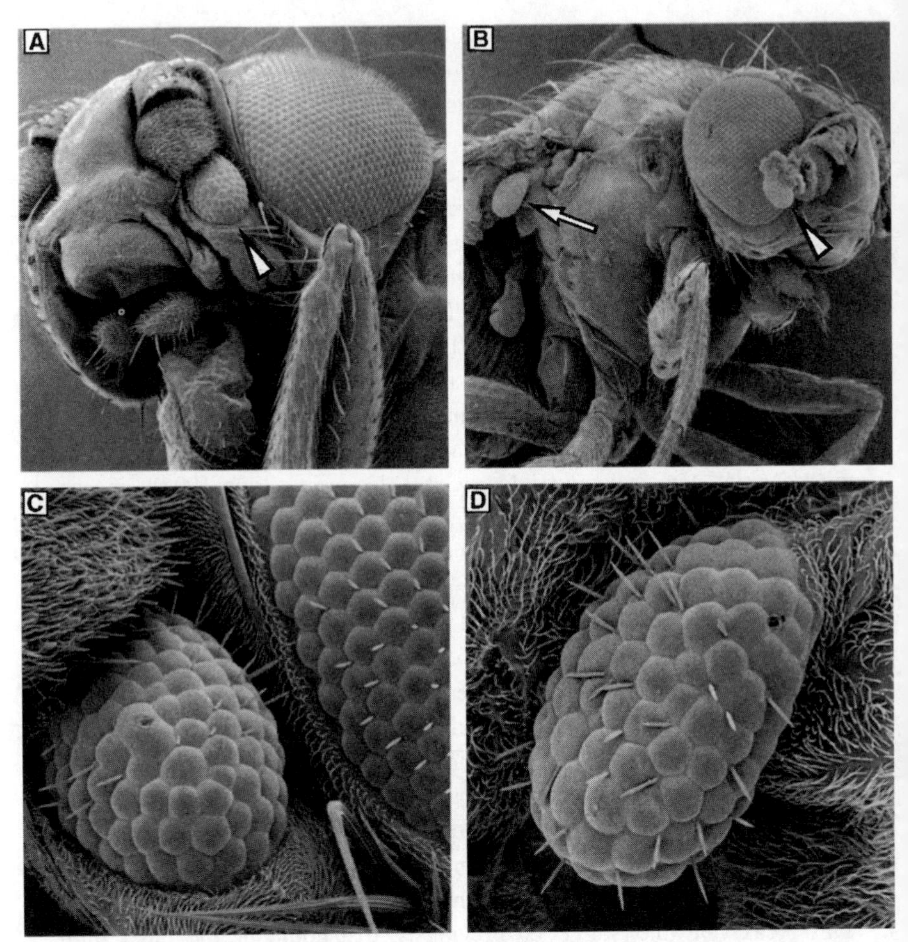


Fig. 5. Scanning electron micrographs of ectopic eyes in *Drosophila*. In B, the arrow points to an ectopic eye under the wing, shown in close-up in D. Such eyes are histologically normal or near normal and are connected to nerves whose axons lead to the brain. The receptors are photosensitive and signals are transmitted by the nerves that innervate them. It is not known whether the eyes are functional or the signals reach the correct domains of the brain. These micrographs are reprinted with permission from Figs. 3B and 3D of G. Halder, P. Callaerts, and W. J. Gehring, "Induction of ectopic eyes by targeted expression of the *eyeless* gene in *Drosophila*," *Science* 267 (1995): 1788–92. Copyright 1995, American Association for the Advancement of Science.

Yet, more remarkably, the mouse protein, expressed in *Drosophila* via genetic engineering, yields *Drosophila* eyes exactly as does the *Drosophila* gene product (Figure 6). Four other *Drosophila* genes are known that, yield eyeless *Drosophila* when appropriately mutated but all of them are expressed

Fig. 6. Scanning electron micrograph with a closeup of an ectopic eye on a *Drosophila* leg. This eye was induced by expression of the product of the mouse *Pax-6* (*small eye*) gene, activated by the same genetic engineering procedures that produced the eyes shown in Fig. 5. But here, the mouse *Pax-6* (*small eye*) gene product replaced the corresponding *Drosophila* gene product. These eyes are virtually indistinguishable from those produced by activation of the *ey* gene as shown in Fig. 5. This micrograph is reprinted with permission from Fig. 5 of G. Halder, P. Callaerts, and W. J. Gehring, "Induction of ectopic eyes by targeted expression of the *eyeless* gene in *Drosophila*," *Science* 267 (1995): 1788–92. Copyright 1995, American Association for the Advancement of Science.

posterior to the *ey* gene in the cascade of phenomena leading to eye formation. Indeed, it is estimated that at least 2,500 genes, and probably more, are activated in the cascade of events that follows expression of the *ey* gene in *Drosophila*. To the authors, these results suggest rather strongly that *ey* is something like a master gene for turning on the eye-making machinery. Such an interpretation is explicitly advanced by Gehring (1994, pp. 7 and 9) for homeotic genes, and applied by Halder, Callaerts, and Gehring (1995) and Quiring et al. (1994) to the *ey* gene. These two papers suggest that the ancestor of the *ey* gene controlled a module for forming primitive eyespots early in metazoan evolution and that the eye-making cascade evolved from this phylogenetically primitive starting point.

## CONFLICTING INTERPRETATIONS

How should we conceive of the eyeless gene in *Drosophila*? Gehring and his colleagues, who showed that *ey* (the *Drosophila* gene) is homologous with mouse and human genes and that the mouse gene can set off the *Drosophila* eye-making cascade, interpret *ey* and its homologues as "a master control gene for eye morphogenesis[16] that is shared by vertebrates and invertebrates" (Halder, Callaerts, and Gehring 1995, p. 1792; see also Quiring et al. 1994). They designate a gene that sets off the cascade of events that changes the identity of a body part and/or produce a specific organ a "control gene." Control genes, including those yielding homeotic mutants, can be detected by the mutant phenotypes they cause. A master control gene is characterized as the first of the genes in the relevant cascade of control genes (Halder, Callaerts, and Gehring 1995, p. 1788). They claim (with all the necessary cautions about the strength of the evidence) that *ey* and its homologues are master control genes.

But, the phrase "master control gene" is controversial – and for good reasons. Gilbert, Opitz, and Raff (1996), for example, would reject any talk of master genes, even in such a dramatic case as this. Their interpretation stems from developmental biology rather than genetics. They employ a modified notion of morphogenetic fields that are laid down by interaction among genes and proteins within bounded domains. These fields are discrete, modular units that respond differently to the activation of a given control gene. They would, therefore, reject the more traditionally genetic interpretation of the Gehring group. The interpretative differences are significant because they enter into the construction and evaluation of hypotheses regarding a wide range of developmental, evolutionary, and genetic phenomena and reflect underlying differences about the ontology in terms of which to interpret organisms.

To illustrate the fundamental character of the issues at stake, consider the claim put forward quite naturally by the Gehring group, that "because [the same master gene] is involved in the genetic control of eye morphogenesis in both mammals and insects, the traditional view [reference to (Plate 1924)] that the vertebrate eye and the compound eye of insects evolved independently has to be reconsidered" (Quiring et al. 1994, p. 785).[17] In other words, Gehring and his colleagues believe that their startling findings show *Drosophila* and mouse eyes to be homologous structures. This revisionist claim apparently rests on the view that structures either *are* or *are not* homologous as wholes, so

[16] At p. 1791, the following rationale is used to support this characterization of the role of *ey* in *Drosophila*: It counts as a master gene "because it can induce ectopic eye structures in at least the imaginal discs of the head and thoracic segments" (see also Quiring et al. 1994, p. 788).

[17] See also p. 788: "If the corresponding genes in flatworms and nemerteans are also involved in eye morphogenesis, the concept that the eyes of invertebrates have evolved completely independently from the vertebrate eye has to be reexamined. Also, the hypothesis that the eye of cephalopods has evolved by convergence with the vertebrate eye is challenged by our recent finding . . . of *Pax-6* – [the mouse homologue of *ey* – RB] related sequences in the squid *Loligo vulgaris*.

that finding a genuine homology deep within the controls for the manufacture of corresponding structures proves that they are homologous. This commitment follows naturally from the view that there is a master gene responsible for making an entire structure and from the finding that the master gene, slightly altered but still functional, is shared between two organisms.[18] In contrast, I argue that both the implicit claim about homology and the explicit characterization of *ey* as a master gene should be rejected.

The objection to master genes can be made in part from within the genetic tradition. The phrase "master gene" *could* be replaced with one that is far less committal – after all, why not call the first gene in such a sequence a "switch gene," say, rather than a master gene? "Switch gene" is a standard genetic term. As is well known, it is commonly used, for example, for genes that control wing pattern in butterflies. Homeotic mutants are those in which a body part (or a portion thereof) has switched identity. Gehring himself (1994, p. 6) mentions as an alternative term "homeotic selector gene." Genes that set off a cascade that alters the identity of a body part or a fundamental pattern or morphology are thus rather naturally characterized as switch or selector genes.

But, is this just trivial semantics? Is there a significant difference between speaking of master genes and speaking of switch genes? I claim that the terminological difference is *not* innocuous. The concept of a master gene carries strong heuristic (not to mention ideological) overtones with it. In particular, it supports claims that lie at the heart of the old divisions between genetics on the one hand and embryology or developmental biology on the other. It supports the metaphor of the genome as a program that contains the entire organism *in potentia*. This is a commitment at the heart of the human-genome program, one that the concept of a switch gene need not support, especially when one recognizes that sometimes the switch in question may be thrown by environmental rather than genetic or other internal changes (cf. seasonal dimorphism in butterflies). We face here all of the familiar issues about the degree to which the so-called information in the genome depends on its cellular and organismic – even its environmental – context. Thus, it is not surprising that many analysts of science and scientists, including many developmental biologists, resist the terminology of master genes (e.g., Keller 1995; Nijhout 1990). I share their resistance and wish to examine further the issue of the role that should be ascribed to genes even in such an extreme case as that of the putative master gene for eye formation.

[18] [Added 2003:] I should have added as a condition for homology that the putative master gene (based on the evidence) is also (probably) derived from a common ancestor.

## HOMOLOGY

To help drive home the biological importance of these conceptual issues, consider the problem of homology. Grant that the Gehring group is right – as it surely is – that the ancestor of the *ey* gene activated a module for forming primitive eyespots early in metazoan evolution and that the eye-making cascades of different lineages evolved from this phylogenetically primitive starting point. How are we to take account of the fact that eyes – and the later parts of the eye-making cascade – evolved so differently in different lineages? Do not the differences in material composition, structure, and morphology between mouse eyes and *Drosophila* eyes resolve the issue in favor of analogy rather than homology? Do not the enormous differences between the genes and gene products recruited in forming mouse and *Drosophila* eyes – and the structural differences in the ways those products are organized force the traditional answer – *analogy!* – on us? But, then, how are we to account for the genetic homology that we just granted? This is a piece of the problem of understanding how to define and apply the concept of homology properly. The topic is too large to pursue at length here; the most that can be hoped for is show that there is a conceptually coherent way of answering these questions in favor of *analogy* rather than *homology*.

For the sake of brevity, I borrow the approach developed by V. Louise Roth (1984)[19] with an addition from Gilbert, Opitz, and Raff (1996). Roth grants that homology requires reproductive community, so that "*genetic relationship* provides the thread which unites all forms of biological homology into one concept, with one definition" (Roth 1984, p. 17). She proposes that homology requires the sharing of a common developmental pathway and points out that this proposal encompasses both serial and special homology. She also emphasizes that *features of structures, rather than the structures themselves, are homologous* and that *the extent of shared differentiated pathways is a matter of degree*. Thus,

> homology is not an all-or-none phenomenon: it is important to recognize *degrees* of homology. The hierarchical nature of phylogenetic homology is widely recognized: as forelimbs, bird wing and bat wing are homologues; as wings they are not – the reptilian common ancestor had forelimbs, not wings. By analogy with phylogenetic relationship, in which taxa can be closely or distantly

[19] [Added 2003:] This representation of Dr. Roth's views was written in ignorance of her work after 1984. Her subsequent articles advance the discussion and remove some of the criticisms raised here. See Roth (1988, 1991, 1994); see also Abouheif et al. (1997) and Roth and Mercer (2000).

related, structures can be biologically homologous to varying degrees – from nearly identical, strongly homologous structures, on down to the very weakest degree of homology, manifested by structures which simply derive from the same germ layer. The points at which developmental paths diverge determine the strength of the homology (Roth 1984, pp. 18–19).

Furthermore, homology must be measured in a way that is appropriate to the structures being examined; different measures of homology are appropriate for genes than are appropriate for limbs, eyes, or behaviors.

> By stressing the genetic basis of homologous relationships I do not mean to imply that one could measure degree of homology in morphological structures most effectively by counting the number of identical codons involved in their developmental process.... While I attribute homology to the genealogy of genes (whereas I would reject a suggestion of homology if structures are created by unrelated sets of genes), the *measure* of homology in morphological structures resides in part at the level of descriptive embryology (Roth 1984, p. 18).

Although Roth's official view is that *features*, not *structures*, are homologous, she does not consistently enforce this position; her account would be greatly improved by so doing. On the resultant account of homology, the fact that certain genes are intimately involved in or even trigger the entire cascade for the formation of, say, the eyes of two different organisms cannot settle the degree of homology between those eyes, for the degree of homology that a comparison will yield depends on the features compared:

> Tying the definition [of homology] to developmental pathways [indeed, more strongly, tying the *degree* of homology to the extent of differentiated developmental pathway held in common – RB] establishes a link between the concept, and the element of continuity (genetic reproduction) at its basis. However, applying it at the level of embryology rather than at the level of genes, in addition to making it more practical, keeps the definition at the organismal level, where the concept was applied originally and where it is most biologically meaningful (Roth 1984, p. 19).

An addition from Gilbert, Opitz, and Raff (1996) helps clarify the point. They make a useful distinction between what they call *process* homology and *structural* homology.[20] The findings regarding segment formation and specification discussed previously show that basic processes underlying segmentation and specification of segment identity in insects and mammals are homologous (and are initiated by genes whose sequence similarity is due to homology) even though the relevant features of the structures produced on

[20] For a defense of the concept of process homologies, see Gilbert and Bolker (2001).

the various segments (e.g., to speak only of insects, wings, and halteres) are not homologous. Putting it more carefully, because homology *is* a matter of degree, the *processes* underlying segment specification and organ formation can be strongly homologous even when the relevant aspects or characteristics of the structures under investigation cannot be counted as homologous. Accordingly, the findings regarding homeotic mutants during the last ten to fifteen years may be summarized as follows: There is now strong support for the immensely surprising result that both segmentation and the production of eyes in vertebrates and invertebrates are initiated by a cascade of homologous processes even though the features of the structures that corresponding segments bear and the eyes of these organisms are, at best, analogous. Thus, despite the strong homology between the series of process control systems in invertebrates and vertebrates that initiate eye formation, nearly any suitable characteristic of the vertebrate and invertebrate structures on corresponding segments should count as analogous rather than homologous. The great surprise at stake is the deep and unsuspected homology of major aspects of the processes and genes underlying the formation of the structures in question, not the homology of the resultant structures. Biologists will be coping with the ramifications of this surprise for a long time to come.

## EXPLANATORY APPARATUS

I take one of the morals of this story to be that we do not yet know what is required for a deep improvement of the explanatory apparatus with which to explain many aspects of organismic development. It is clear that many incredibly intricate multileveled domains, mechanisms, processes, structures, and so on enter into development. Furthermore, many of these are formed (pardon the pun) "on the fly" – that is, they are not laid out in advance but rather arise in interactions between genes and proteins that come to form a rapidly shifting tartan of boundaries between domains and define something like morphological fields in the midst of the ongoing processes of cell-type specification, tissue formation, organogenesis, and so on. At any stage of development, some of the relevant modules that enter into normal development preexist, others are formed in the course of events, and others will or will not be formed according to the status and condition of interacting units and modules at key moments in the processes in question. The shuffling of process modules and alterations of their interactions is a form of evolutionary tinkering. Recognition of this topic opens up a wide range of interesting investigations, which is at its very beginning.

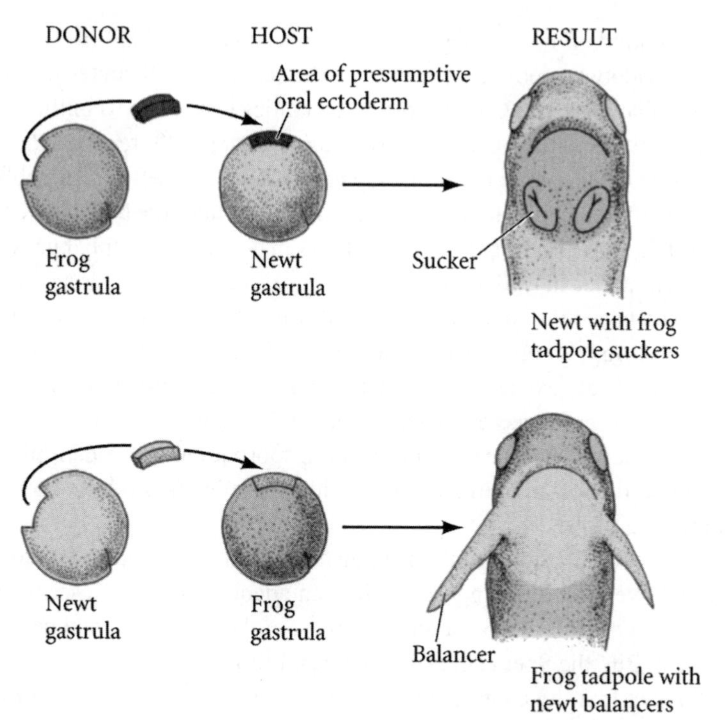


Fig. 7. Reciprocal transplantation between newt and frog gastrulae. In each case, indifferent ectoderm is transplanted into the area of presumptive oral ectoderm of the foreign host. The result is a product that would normally belong near the mouth of the *donor*, not the host, species. (The figure shows newt larvae with tadpole suckers and tadpoles with newt balancers.) Such findings show the genetic specificity of the donor material and its responsiveness to the control system (morphological field?) that determines the fate of the region into which it is transplanted. This figure was redrawn after Hamburgh (1970), based on experiments first reported in Spemann and Schotté (1932). Reproduced, with permission, of Sinauer Associates and Scott F. Gilbert, *Developmental Biology*, 7th ed., Sunderland, MA: Sinauer, 2003. I am grateful to Prof. Gilbert for supplying an electronic copy of his figure.

I mention two final examples, both of xenoplastic transplantation, to highlight the point. More than seventy years ago, Spemann and Schotté (1932) showed that reciprocal transplants between presumptive flank ectoderm in early frog and newt gastrulae, placed in the ectodermal region destined to become the mouth in the other animal's gastrula, yielded parts related to the mouth. But, the frog material in the newt yielded tadpole suckers, whereas the newt material in the frog yielded newt balancers (Figure 7). Harrison (1933, p. 318) reports that Spemann described this result as follows: "The ectoderm says to the inducer, 'You tell me to make a mouth; all right, I'll do so, but I can't make your kind of mouth; I can make my own and I'll do that'."[21] This result has been widely reproduced. In contrast, the second is still contested. But, it is remarkable in seeming to recover potentialities lost about 10 million years ago. Some twenty-four years ago, Kollar and Fisher (1980) reported that they had activated the machinery for making nonmammalian teeth, presumably the teeth of a proximal ancestor of chickens, in chick epithelium transplanted into mice. If this result is really reliable, it shows that there are deeply buried processes that, under ill-understood circumstances, can be activated and recruit determined tissue into quite different pathways. I interpret these distinct results as showing not only that morphological fields are changing entities but also that their very existence (if we take them to be causally efficacious) is at best relative to processes and events that we do not understand. We have no reason to believe that there is a tooth-forming morphological field in chickens; yet, under rather bizarre circumstances, chicken cells can produce or respond to something of the sort.

I am skeptical (though my skepticism is not an argument) whether, at this point, we have anything resembling an adequate way of describing the multileveled causally relevant units for most of the developmental processes of interest. But, the Spemann and Schotté and Kollar and Fisher results illustrate the fact that the controls of such processes as making mouths or teeth are preserved across species barriers, whereas the details of the product are controlled at or close to the level of the species. On grounds such as these, I am as skeptical of transformations of concepts taken from developmental biology (e.g., the concept of morphogenetic fields) as I am of restricting ourselves to narrowly genetic concepts. Until we know how to delimit adequately the boundaries of the causally relevant units in a developmental network, to determine when they are brought into existence or made causally efficacious, and to detail the mechanisms that yield different results when different materials from different sources are placed within the boundaries of a given unit, we will not be able to assess adequately the value of any attempted integrative explanatory apparatus for the phenomena of development. And because, to repeat, the interactions in question involve multileveled processes and sometimes evanescent modules, made by bootstrapping "on the fly" and effective only in interaction with other such modules, we are far from understanding what exactly it is that we are talking about.

This skeptical stance is meant as a spur, not a barrier, to research. The challenge is to articulate the strengths and limitations of various approaches attempting to integrate genetics with our current understanding of

[21] Scott Gilbert kindly called this example to my attention. See Gilbert (1988, pp. 561–2).

developmental phenomena and to enforce contact between them as best we can. More especially, the deep challenge is to work out how to demarcate the relevant boundaries of the many interactive and dynamic processes required to cause the formation of particular structures so that one can balance the need for identifying the relevant units in a causal analysis with the dynamic and multileveled character of the processes under investigation. This is a task that will require fascinating conceptual as well as empirical research – a task I challenge the reader to take up.

## REFERENCES

Abouheif, E., M. Akam, W. J. Dickenson, P. W. H. Holland, A. Meyer, N. H. Patel, R. A. Raff, V. L. Roth, and G. A. Wray. 1997. "Homology and developmental genes." *Trends in Genetics* 13: 432–3.

Allen, G. E. 1986. "T. H. Morgan and the split between embryology and genetics, 1910–1935." In *A History of Embryology*, eds. T. J. Horder, J. A. Witkowski, and C. C. Wylie. Cambridge: Cambridge University Press, 113–46.

Arendt, D., and K. Nübler-Jung. 1994. "Inversion of dorso-ventral axis?" *Nature* 371: 26.

Bürglin, T. F. 1994. "A comprehensive classification of the homeobox genes." In *Guidebook to the Homeobox Genes*, ed. D. Duboule. Oxford: Oxford University Press.

de Robertis, E. M. 1994. "The homeobox in cell differentiation and evolution." In *Guidebook to the Homeobox Genes*, ed. D. Duboule. Oxford: Oxford University Press, 11–23.

de Robertis, E. M., E. A. Mortita, and K. W. Y. Cho. 1991. "Gradient fields and homeobox genes." *Development* 112: 669–78.

Gehring, W. J. 1994. "A history of the homeobox." In *Guidebook to the Homeobox Genes*, ed. D. Duboule. Oxford: Oxford University Press, 1–10.

Gehring, W. J. 1998. *Master Control Genes in Development and Evolution: The Homeobox Story*. New Haven: Yale University Press.

Gilbert, S. F. 1980. "Owen's vertebral archetype and evolutionary genetics – a Platonic appreciation." *Perspectives in Biology and Medicine* 23: 475–88.

Gilbert, S. F. 1988, 2nd ed. *Developmental Biology*. Sunderland, MA: Sinauer.

Gilbert, S. F., ed. 1991. *A Conceptual History of Modern Embryology*. New York: Plenum.

Gilbert, S. F. 2003. *Developmental Biology*, 7th ed. Sunderland, MA: Sinauer.

Gilbert, S. F., and J. Bolker. 2001. "Homologies of process and modular elements of embryonic construction." *Journal of Experimental Zoology (Molecular and Developmental Evolution)* 291: 1–12.

Gilbert, S. F., J. M. Opitz, and R. A. Raff. 1996. "Resynthesizing evolutionary and developmental biology." *Developmental Biology* 173: 357–72.

Halder, G., P. Callaerts, and W. J. Gehring. 1995. "Induction of ectopic eyes by targeted expression of the eyeless gene in *Drosophila*." *Science* 267, no. 5205: 1788–92.

Hamburgh, M. 1970. *Theories of Differentiation*. New York: Elsevier.

Harrison, R. G. 1933. "Some difficulties of the determination problem." *American Naturalist* 67: 306–21.

Harwood, J. H. 1984. "History of genetics in Germany." *The Mendel Newsletter* 24: 3.

Harwood, J. H. 1993. *Styles of Scientific Thought: The German Genetics Community, 1990–1933*. Chicago and London: University of Chicago Press.

Horder, T. J., J. A. Witkowski, and C. C. Wylie, eds. 1986. *A History of Embryology*. Cambridge: Cambridge University Press.

Keller, E. F. 1995. *Refiguring Life*. New York: Columbia University Press.

Kollar, E. J., and C. Fisher. 1980. "Tooth induction in chick epithelium: Expression of quiescent genes for enamel synthesis." *Science* 207: 993–5.

Lawrence, P. A. 1992. *The Making of a Fly: The Genetics of Animal Design*. Oxford: Blackwell Scientific Publications.

Lewin, B. 1990. *Genes IV*. Oxford: Oxford University Press.

Maienschein, J. 1986. "Preformation or new formation – or neither or both?" In *A History of Embryology*, eds. T. J. Horder, J. A. Witkowski, and C. C. Wylie. Cambridge: Cambridge University Press, 73–108.

Malicki, J., C. Cianetti, C. Peschle, and W. McGinnis. 1992. "Human HOX4B regulatory element provides head-specific expression in *Drosophila* embryos." *Nature* 358: 345–7.

McGinnis, N., M. Kuziora, and W. McGinnis. 1990. "Human Hox 4.2 and *Drosophila* deformed encode similar regulatory specificities in *Drosophila* embryos and larvae." *Cell* 63: 969–76.

Morgan, T. H. 1934. *Embryology and Genetics*. New York: Columbia University Press.

Morgan, T. H. 1935. "The relation of genetics to physiology and medicine (Nobel Lecture, Stockholm, June 4, 1934)." *Scientific Monthly* 41: 5–18.

Nijhout, H. F. 1990. "Metaphors and the role of genes in development." *BioEssays* 12: 441–6.

Opitz, J. M. 1985. "Editorial comment: The developmental field concept." *American Journal of Medical Genetics* 21: 1–11.

Panchen, A. L. 2001. "Étienne Geoffroy St.-Hilaire: Father of 'evo-devo'?" *Evolution and Development* 3: 41–9.

Plate, L. 1924. *Allgemeine Zoologie und Abstammungslehre*. Jena: Fischer.

Quiring, R., U. Walldorf, U. Kloter, and W. J. Gehring. 1994. "Homology of the eyeless gene of *Drosophila* to the *small eye* gene in mice and *Aniridia* in humans." *Science* 265: 785–9.

Raff, R. A., and T. C. Kaufman. 1983. *Embryos, Genes, and Evolution: The Developmental-Genetic Basis of Evolutionary Change*. New York, Macmillan.

Roth, V. L. 1984. "On homology." *Biological Journal of the Linnaean Society* 22: 13–29.

Roth, V. L. 1988. "The biological basis of homology." In *Ontogeny and Systematics*, ed. C. J. Humphries. New York: Columbia University Press, 1–26.

Roth, V. L. 1991. "Homology and hierarchies: Problems solved and unresolved." In *Journal of Evolutionary Biology*, 167–94.

Roth, V. L. 1994. "Within and between organisms: Replicators, lineages, and homologues." In *Homology: The Hierarchical Basis of Comparative Biology*, ed. B. K. Hall. San Diego and New York: Academic Press, 301–37.

Roth, V. L., and J. M. Mercer. 2000. "Morphometrics in development and evolution." *American Zoologist* 40: 801–10.

Sander, K. 1986. "The role of genes in ontogenesis – evolving concepts from 1883–1983 as perceived by an insect embryologist." In *A History of Embryology*, eds. T. J. Horder, J. A. Witkowski, and C. C. Wylie. Cambridge: Cambridge University Press, 363–95.
Sander, K. 1994. "Of gradients and genes: Developmental concepts of Theodor Boveri and his students." In *Roux's Archives of Developmental Biology*, 203: 295–7.
Sapp, J. 1983. "The struggle for authority in the field of heredity, 1900–1932: New perspectives on the rise of genetics." *Journal of the History of Biology* 16: 311–42.
Sapp, J. 1986. "Inside the cell: Genetic methodology and the case of the cytoplasm." In *The Politics and Rhetoric of Scientific Method*, eds. J. A. Schuster and R. R. Yeo. Dordrecht, Holland: D. Reidel, 167–202.
Sapp, J. 1987. *Beyond the Gene: Cytoplasmic Inheritance and the Struggle for Authority in Genetics*. New York: Oxford University Press.
Spemann, H., and O. Schotté. 1932. "Über xenoplastische Transplantation als Mittel zur Analyse der embryonalen Induction." *Naturwissenschaften* 20: 463–7.
Wolpert, L. 1986. "Gradients, position, and pattern: A history." In *A History of Embryology*, eds. T. J. Horder, J. A. Witkowski, and C. C. Wylie. Cambridge: Cambridge University Press, 347–62.

# 12

# Reconceiving Animals and Their Evolution

## On Some Consequences of New Research on the Modularity of Development and Evolution[1]

The last fifteen years have seen an ongoing synthesis among developmental biology, evolutionary biology, and molecular genetics.[2] A new discipline, "evolutionary developmental biology," is forcing biologists to reconceive evolutionary history, evolutionary processes, and the ways in which animals are constructed. In this chapter, I examine some work bearing on how animals (including humans) are put together. A key claim is that evolution deploys ancient modular processes and tinkers with multileveled modular parts, many also ancient, yielding organisms whose relationships – because of modular construction – are far more complex and interesting than had been suspected until recently. For example, all segmented animals share regulatory machinery that demarcates and specifies identities of body segments and switches on the formation of some organs (e.g., eyes). Some processes, bits of machinery, and parts are recycled and reused repeatedly, both in evolution and in the development of a single animal. Such claims require major rethinking

[1] Earlier versions of this chapter were presented at the International Society for History, Philosophy, and Social Studies of Biology, at Ohio University in 2001, and at a Colloquium sponsored by the Virginia Tech Department of Philosophy and Center for Science and Technology Studies in 2002. I am grateful to all three audiences and to numerous colleagues for comments on this and a companion presentation. Colleagues deserving particular thanks are Bill FitzPatrick, Frietson Galis, Scott Gilbert, Marjorie Grene, Paul Siegel, Günter Wagner, and Lee Zwanziger. Scott Gilbert generously supplied electronic copies of eight of the figures in this chapter. Thanks, Scott!

[2] The best single resource for the biology that goes into the new discipline is the last chapter of either of the last two editions of Gilbert's *Developmental Biology* (Gilbert 2000, 2003). Gilbert's book does an extraordinary job of covering the substantial content of present-day developmental biology. This chapter was built using the sixth edition because the seventh arrived too recently for me to consider revisions. Other important books written in the last fifteen years include Arthur (1988, 1997); Carroll, Grenier, and Weatherbee (2001); Davidson (2001); Gerhart and Kirschner (1997); Hall (1992); Hall (1994); Hall (1999); Hall and Olson (2003); Jablonka and Lamb (1995); Keller (2003); Müller and Newman (2003); Pigliucci (2001); Raff (1996); Schlichting and Pigliucci (1998); Wagner (2001); and Wilkins (2002).

of how animals are put together and raise issues about how – and the extent to which – animals are harmoniously integrated. Our understanding of how synchronic and sequential developmental processes are controlled to yield an organism is still far from complete. We do not yet understand the philosophical implications of this new work, but I suggest that they include a limited, nonvitalist form of holism.

## INTRODUCTION

The task of this chapter is somewhat complicated. I devote most of it to some philosophically interesting developments in evolutionary developmental biology and then begin to extract from them some issues they raise about the nature of animals. To do this, I must cover a great deal of biology rather sketchily. I try to provide enough exposition of the technicalities to raise the most important issues calling for clarification and substantive investigation.

To start, I show that animals are put together differently than we imagined until recently. What follows are a few salient points. The genome as such does not constitute a blueprint for the organism; rather, a variety of control systems are involved. Some of these are deployed in eggs before fertilization. The control systems are crucial in dividing animals into segments and, in a few known cases, they initiate the formation of particular organs (e.g., brain, eyes, heart, and kidneys). The operation of these control systems are typically staged, coordinated, or synchronized by exogenous, hormonal, and other signals. A great deal of self-assembly is involved; many of the control systems and organs are built in the middle of the process of development, combining resources found in the egg and sperm (including the genome) with signals provided by the environment and the cellular context and location. Although copies of the control systems are built during development, many of the controls deployed in building animal bodies are very ancient – they are indirect copies of controls that already existed before the lineages that yielded insects and vertebrates diverged (i.e., they are well over 300 million years old). The control systems that regulate gene expression are exquisitely context-sensitive and have been recycled – both on an evolutionary scale and within the development of a single organism – to accomplish a great variety of tasks.

We do not yet understand how the actions of these control systems are integrated to yield the very different organisms that fall within these enormous lineages. The organization of the genome has a major role in this integration, a point that I touch on later. Nonetheless, it is open to question whether there is a central control unit of any sort responsible for the integration of the organism. It is just possible that a series of self-assembly units accomplish much of what goes into making an animal – that is, that animals are put together as the result of cascades of self-assembly processes that run without much informational input from any sort of central control unit. And, within any animal, there are parts with conflicting interests that must have reached some sort of stable but imperfect equilibrium. If all of this is correct, one of the key problems of developmental biology – potentially tractable to the new molecular and computational methods – will be how integration of the organism is achieved in development. In any event, however this is done, animals have been cobbled together by evolutionary tinkering, in which the reshuffling of miscellaneous pieces and what I call control modules is at least as important as genetic novelties.

Most of the developments I discuss are quite recent. The biological findings are rather specialized and have been obtained largely thanks to the power of new technologies. Accordingly, they are not yet widely known to nonspecialists. For this reason, I devote most of the chapter to an overview of some findings of interest. Only then can I take on, all too briefly, some of the larger issues that they raise.

## SOME BASIC BIOLOGICAL SURPRISES REGARDING THE REGULATION OF DEVELOPMENT

Our topic is the regulation of development in multicellular animals. For ease of exposition, I discuss only segmented animals, using examples from insects (primarily the fruit fly, *Drosophila*, whose segmentation is fairly well understood) and mammals (primarily rats and mice, secondarily humans). As will soon become clear, the results discussed here apply to a much wider range of organisms than anyone expected as recently as fifteen years ago. These results could not have been obtained without use of new biological technologies, some of which have been developed in the Human Genome Project. I hope to convince you of the power and importance of a few key findings.

To begin, consider one of the more startling claims made by the teams that mapped the human genome (International Genome Sequencing Consortium 2001; Venter et al. 2001; and other articles in the same journal issues). Although these numbers are still contentious (Hogenesch et al. 2001; Wright et al. 2001) using conventional molecular means for delimiting genes, human beings seem to have only about 26,000 to 38,000 genes (probably 30,000 to

35,000), which is approximately one third of the number predicted about a decade ago and less than double the number identified in *Drosophila*. This surprisingly low number of genes is counteracted, however, by a corresponding complexity of the regulatory apparatus that makes far more than 200,000 proteins from these roughly 30,000 genes – more than 1,200,000 if one includes antibody proteins. So, *there is no such thing as a 1:1 relationship between genes and the proteins they produce*.

A second point: Depending on how one counts and on some uncertainties about the interpretation of the molecular findings, *only about 1.5 percent to at most 4 percent of the DNA in our genome codes for proteins*. A significant portion, possibly about half of the rest, is highly repetitive DNA (sometimes called selfish DNA), much of which has no known function. *But something like 40 percent of the genome has regulatory significance* – that is, it includes material that in one way or another makes, serves as, modulates, or interacts with signals affecting the transcription of DNA or the processing of RNA made from that DNA.

Many mechanisms intervene between the transcription of protein-coding sequences to RNA and the eventual production of a protein. For simplicity, I concentrate on only two mechanisms, although there are at least two dozen of them. The two I discuss involve post-transcriptional modification of the sequences transcribed from DNA to RNA. To facilitate the discussion, I use Gilbert's simplified diagram of a real case, illustrating some key steps involved in producing $\beta$-globin and hemoglobin (Figure 1).[3] The stretch of DNA that encodes the sequence information for the protein begins with a promoter region, required for making an RNA transcript, ultimately processed into the messenger RNA that specifies the sequence of amino acids in the protein. As the first step in the diagram shows, the DNA in question is transcribed to a shorter length of RNA, beginning with a so-called leader sequence and ending with a tail composed of a poly-A string (i.e., repeated adenines attached to the sugar chain of the RNA). This nuclear RNA, however, contains "introns" – intervening sequences that must be excised, with the remaining RNA spliced at the excision joints – if the resulting ("mature") messenger RNA (mRNA) is to pass through a nuclear pore into the cytoplasm.[4] (Some of the excised material forms "small nuclear RNAs," some of which play further roles in the

[3] This and most of the other figures are taken from Gilbert (2003). I am grateful to Prof. Gilbert for encouraging me to make this use of his work.

[4] One of the many hypotheses about the evolutionary reason for this constraint on getting the RNA through the nuclear pore is that the processing involved helps protect against RNAs made by retroviruses from getting out into the cytoplasm and doing damage to cells and the organisms that they have infected.

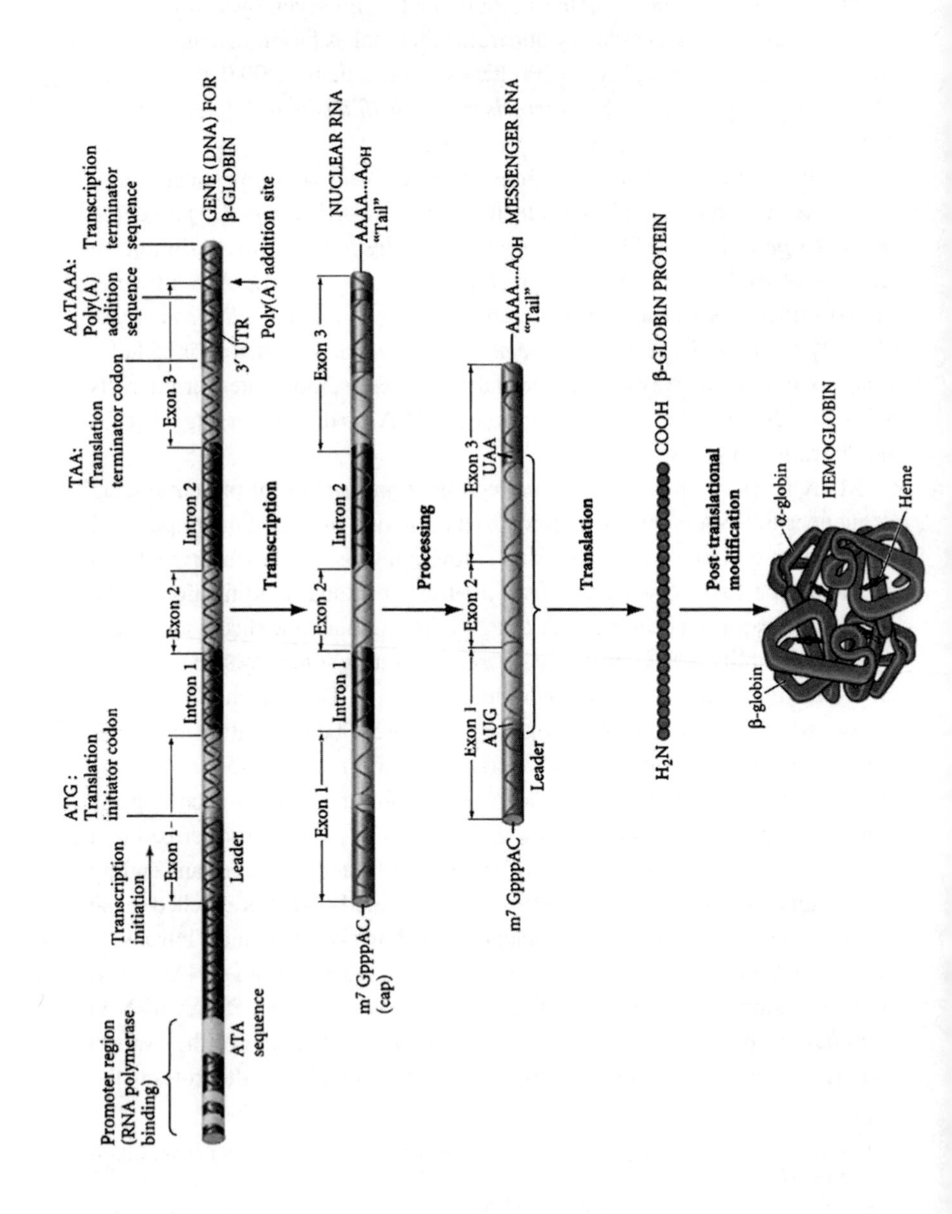
Promoter region
(RNA polymerase
binding)
ATA
sequence
Transcription
initiation
Exon 1
Leader
ATG:
Translation
initiator codon
Intron 1
Exon 2
Intron 2
TAA:
Translation
terminator codon
Exon 3
AATAAA:
Poly(A)
addition
sequence
3′ UTR
Poly(A) addition site
Transcription
terminator
sequence
GENE (DNA) FOR
β-GLOBIN
Transcription
m7 GpppAC
(cap)
Exon 1
Intron 1
Exon 2
Intron 2
Exon 3
AAAA...AOH
"Tail"
NUCLEAR RNA
Processing
m7 GpppAC
Exon 1
AUG
Leader
Exon 2
Exon 3
UAA
AAAA...AOH
"Tail"
MESSENGER RNA
Translation
H2N
COOH
β-GLOBIN PROTEIN
Post-translational
modification
β-globin
α-globin
Heme
HEMOGLOBIN

processing of RNAs in the nucleus.) The mature mRNA contains a shorter tail, none of the introns, and a signal (the sequence AUG) that marks the end of the leader and the beginning of the RNA that will be translated in making the $\beta$-globin protein. This protein, in turn, may go through "post-translational processing" in which it is altered in various ways and combined with some other similarly produced products to make a hemoglobin molecule.

The excision of the introns – noncoding segments of the raw RNA transcript[5] – joins together previously separated coding segments (called exons) that are now contiguous in the processed RNA transcript. The joined pieces contain the actual signal for the sequence of amino acids produced on ribosomes in the cytoplasm. Virtually all protein-producing genes in animals have from one to dozens of introns. The gene for $\beta$-globin is simpler than average in this respect – it has only two introns. Nearly all the DNA that codes for proteins in plants and animals is not contiguous on the DNA molecule. So, point three: *One reason that DNA sequence does not correspond exactly with the protein product of an animal gene is that the signal for the amino-acid sequence of the protein is typically put together out of disjoint pieces of DNA*. Thus, the sequence of DNA nucleotides *within* genes does not correspond neatly with the sequence of amino acids in the proteins that our bodies make. Genes are made of modular pieces, some of which correspond to particular domains performing particular functions in the ultimate protein product – and in evolutionary history many of them have been duplicated and/or shuffled around from one gene to another. So, subgenic modular units in the DNA are important in building proteins and in evolution – pieces of genes are put together in building proteins.

The second mechanism to be discussed is alternative splicing of RNA transcripts (Figure 2). Different cells utilize variants of closely related proteins. Some protein domains perform particular functions – for example, binding

[5] Rarely, introns actually contain coding material that belongs to another gene. Thus, strictly speaking, they need not be noncoding, but if they contain coding segments, those segments are part of the code for an entirely different protein.

---

Fig. 1. A schematic summary of steps involved in producing $\beta$-globin and hemoglobin. Top: DNA. Second: Transcription yields nuclear RNA with a cap, exons, introns, and untranslated regions at both ends. Third: Removal of introns in the nucleus yields mRNA. Fourth: Translation yields the polypeptide (string of amino acids) of $\beta$-globin. Final: Post-translational processing may join (and modify) two $\beta$-globins and two $\alpha$-globins, configured with four hemes to yield hemoglobin. Reproduced with permission of Sinauer Associates and Prof. Gilbert (2003, Fig. 5.3). I am grateful to Prof. Gilbert for supplying me with electronic copies of his figures.

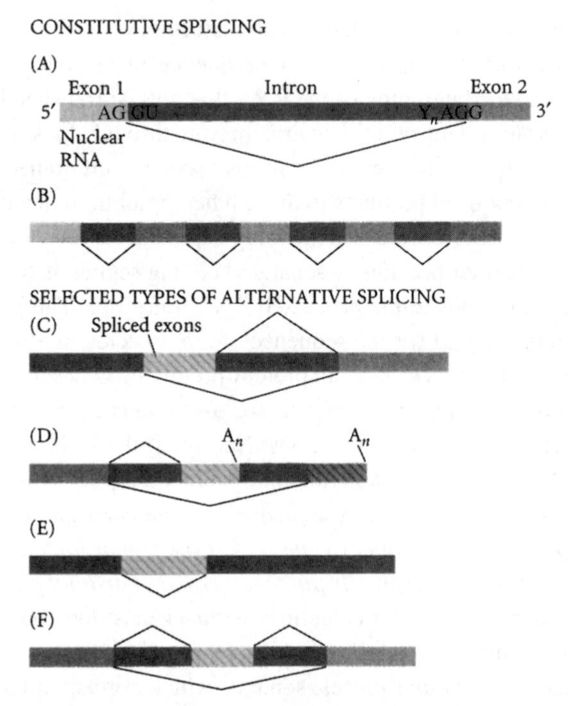


Fig. 2. Schematic diagram of alternative splicing of nuclear RNA. Reproduced with permission of Sinauer Associates and Prof. Gilbert (2003, Fig. 5.27), after Horowitz and Kraianer (1995).

the rest of the protein to some particular sort of cell membrane. Other domains perform different functions – for example, in the case of an immune molecule, recognizing a particular antigen. If what counts as genes are stretches of DNA that produce raw RNA transcripts like those illustrated in the image of the hemoglobin gene, then many animal genes yield RNA that is spliced together differently in different circumstances or in different cells. Figure 3 presents a schematic diagram of a real case: the rat $\alpha$-tropomyosin gene, which is regulated so that it yields different proteins in different cellular contexts by means of alternative splicing. So, the very same gene can yield quite different proteins, often by putting alternative modular pieces performing particular functions together. This process often yields closely related proteins – for example, a soluble immune protein that recognizes a particular antigen and a membrane-bound protein that recognizes the same antigen. In the course of evolution, exons can be duplicated or shuffled so that the signal coding

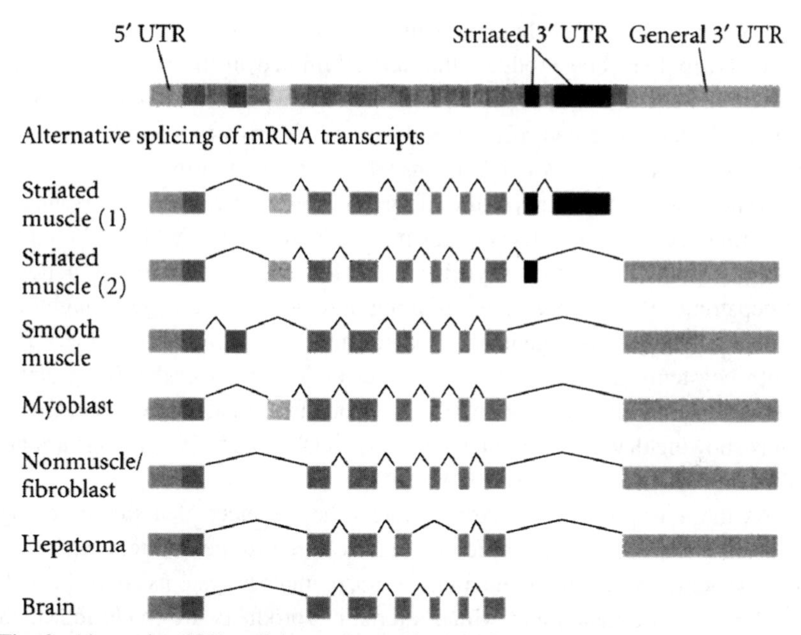


Fig. 3. Alternative RNA splicing to form a family of rat $\alpha$-tropomyosin proteins. The DNA sequence is represented at the top. Thin lines represent sequences that become introns and are spliced out in forming mature mRNA. Reproduced with permission of Sinauer Associates and Prof. Gilbert (2003, Fig. 5.28), after Breitbart, Andreadis, and Nadal-Ginard (1987).

for a domain performing a particular function can be spliced into an entirely different protein.[6]

About 40 percent of human genes are known to be alternatively spliced. Alternative splicing must be highly controlled so that the "right" protein ends up, by and large, being produced in the right place at the right time. However, our understanding of the system of controls that do this job is still quite poor. Suffice it to say that those controls are extremely complex. They often involve several layers of double and triple assurance, inhibitors that block misexpression, inhibitors of those inhibitors to allow expression, and so on. These devices affect whether, when, and where the product is made; which of the alternative products is made; the rate at which the product is synthesized; and its total quantity. In short, gene expression is, in general,

[6] Thus, another evolutionary advantage obtained from the modular exon-intron structure is that it makes it possible to move domains independently of one another, combining segments of proteins that serve distinct functions without having to build novel proteins from scratch.

tightly controlled – and the agents that do the controlling typically are not genes but rather gene products that are assembled in the relevant cells or tissues into control devices – devices that are sensitive to internal and external signals. When those controls break down, all kinds of problems arise. One result may be cancer – which happens when some part of the control system malfunctions so that manufacture of one or more particular products and/or one or more cell types is not shut down when it normally would be. (You may have noticed that the $\alpha$-tropomyosin gene makes products that are components of hepatomas [liver cancers] and of brain tissue – so that this gene ought not be defined simply as a gene for striated muscle or as a cancer gene.) The control systems are quite baroque, as is clear from the fact that some genes produce an enormous number of distinct proteins. Keller, for instance, cites a study showing that one particular gene may, in fact, yield 576 splicing variants (Black 1998, cited by Keller 2000).

As the $\alpha$-tropomyosin case illustrates, the products of a single coding region are often quite distinct proteins that are used in quite different contexts – such as brain, liver, and skeletal muscle. So, regulatory systems control which products a gene yields and which alternative products are made in which circumstances. The success or failure of those systems in controlling what is made, in what circumstances, by particular coding regions ("genes" as counted in the Human Genome Project papers) is crucial in building organisms and required for the well-being of those organisms.

Indeed, a consensus has already emerged that the unexpectedly low increase in the number of genes in mammals (including humans) in comparison with much simpler organisms means that differences in the genes, as such, cannot account for our biological differences from other animals. Rather, the differences between mammals and insects, or among primates, rest primarily on the regulation of gene expression and on what happens in the construction of the organism after the first genes are expressed in the embryo. Differences in the regulatory apparatus that determine which genes are used, when, and how are at least as important as the differences in the coding segments of the genes in question. Thus, in their triumphal article announcing the draft sequence of the human genome, Venter et al. (2001) argued that the reductionist paradigm that drove molecular biology – and much else in biology – is seriously deficient (on this point, see also Laubichler 2000). It is impossible to determine from the exact sequence of nucleotides in DNA the function(s) of many, perhaps most, genes. And the ambiguities caused by alternative splicing mean that the genes in question typically do not have a unique protein-encoding structure and do not encode a unique protein. It follows, I believe, that genes alone do not suffice to provide a blueprint for the construction of an animal.

Rather, regulatory systems, integrated in ways that are not yet understood, but not wholly specified by the genes, are at the heart of the construction of an organism.

This dramatic departure from genetic reductionism provides one of the major conclusions of this chapter and of the entire book. *Genes alone do not provide a blueprint for the organism.* During the last half of the twentieth century, reductionism, specifically genetic reductionism, rode high in biology. *During the next few decades, I believe, biologists will highlight the roles played in constructing organisms by dynamic regulatory systems above the level of the genome.* The result will be a nonvitalist but much more holistic, vision of the organism, one that places the integration of the organism at the focus of attention. In short, our new understanding of the apparatus regulating gene expression has undermined classical genetic determinism. The remainder of this chapter makes this claim clearer and amplifies some of its ramifications.

## ON BUILDING AN ORGANISM BY TINKERING

The next biological topic concerns some findings about how body segments are laid down, how segment identities are determined, and how some organs are generated. I argue for a sharp separation between the processes that regulate development and the products of those processes. There are strong indications that homologous segments can be formed by distinct processes, that homologous processes can yield utterly distinct products, and that homologous processes can yield organs that perform similar functions but do not have homologous properties. It will require one more round of expository work to explain and justify these fundamental claims.

The biological findings also support the claim that animals are cobbled together by a relatively small number of processes that are very highly conserved and that draw on relatively standard materials, many of which are found in most animals. As early as 1977, Jacob suggested that evolution is a tinkerer – a "bricoleur," to use the more descriptive French term (Jacob 1977, 1982; see also Duboule and Wilkins 1998). For almost twenty years, most evolutionists treated this as a pretty metaphor but paid it little heed. By now, however, this image seems to be exactly right. In the course of evolution, animals are put together in miscellaneous ways, by jerry-built combinations of ancient processes, coopting whatever components are at hand. Evolution *is* a tinkerer.

Much of the relevant work was done on the fruit fly *Drosophila*. About fifteen years ago, new molecular tools made it possible to visualize the products of many of the genes whose products act during development. Thanks to these tools and the immense store of knowledge of *Drosophila* genetics, *Drosophila* quickly became the best-explored insect in developmental biology and a workhorse for developmental genetics.

I explain briefly what I mean by homologous segments and then turn to the regulation of developmental genes. There are two kinds of homology to consider. The first is serial homology, the second homology by descent. Figure 3 in Chapter 11 (redrawn from Raff and Kaufman 1983) diagrams a hypothetical evolutionary lineage, from annelids to pterygote insects. It reminds us of a number of points. Homology is a matter of degree. The segments of the first animal shown are clearly serially homologous. They probably originated from duplication of segments – a phenomenon I illustrate shortly. Considered just as segments, all the segments of insects are serially homologous in that they are descended from serially homologous segments. But, because the segments acquired more specific identities in the course of evolution, finer-grained homologies arose. Thus, the order *Insecta* arose in this lineage when the three thoracic segments, each bearing a pair of legs, became distinct from the others and became the only leg-bearing segments. The fact that all insects have only six legs,[7] all belonging by descent to these segments shows that the limitation to three thoracic leg-bearing segments is, for some reason, deeply entrenched and that this homology is of considerable importance to what an insect is and how it is built. The segments of the last insect in this lineage are all homologous by descent to segments of the most remote ancestor shown. The contrast between thoracic segments and abdominal segments is due to the descent of these insects from an ancestor that had locked in a developmental system that yields only three thoracic segments and restricted leg-making to those segments. Similar things apply, of course, to vertebrates, as is shown by the fact that a bat wing is homologous *as a forelimb* to a cow's leg, an orangutan's arm, and a human's arm, but only the latter two are homologous *as arms*.

Let me now tie all this to the new work in development by reference to one of the most famous homeotic mutants of *Drosophila*. The wild type has one pair of wings on its second thoracic segment and small balancing organs called halteres on its third thoracic segment. The mutant, called bithorax (described briefly in Chapter 11), is a striking and unusual case of serial homology (Figure 4). Two mutations have converted what should be the third thoracic

[7] The fact that leg formation is sometimes blocked does not alter the point; only three segments have leg-making apparatus or the triggers for making legs, whether or not the production of legs is blocked in some way.

Fig. 4. Four-winged fly, produced by combining *bithorax* and *postbithorax* mutations. Reproduced with permission from Lawrence (1992, Plate 5.1).

segment of this *Drosophila* into a second thoracic segment. Thus, the fly has an extra second thoracic segment and, accordingly, an extra pair of wings, no third thoracic segment, and thus no halteres.

At the end of the nineteenth century, William Bateson coined the term "homeosis" for the phenomenon of having a segment or part with the wrong identity; early in the twentieth century, he named a mutation that yielded homeosis a homeotic mutation. So, this fly has a homeotic mutation, which we now know took place in a complex of three genes called "ultrabithorax."

So-called hox genes, discussed briefly in the previous chapter, often produce homeotic mutants. They all possess a homeobox, which encodes a sixty-amino-acid-long homeodomain in the corresponding protein. These protein domains are highly conserved in *all* animals (and even in yeasts), and they attach to DNA at highly specific sites. Proteins with homeodomains regulate the transcription and expression of genes downstream from those attachment sites, most often by turning on transcription, sometimes by affecting its rate or turning it off. In short, hox genes produce proteins that alter the transcription of specific regions of DNA. Those proteins are *signal transduction factors*; they are components of *signal transduction modules*. These modules are composed of interacting proteins (and signals that encode those proteins) that act on stretches of DNA. These modules are crucial elements in the regulatory apparatus that controls development and they are typically triggered by the presence, inactivation, or removal of signaling molecules (sometimes environmental, sometimes made by the organism). The mutations that cause

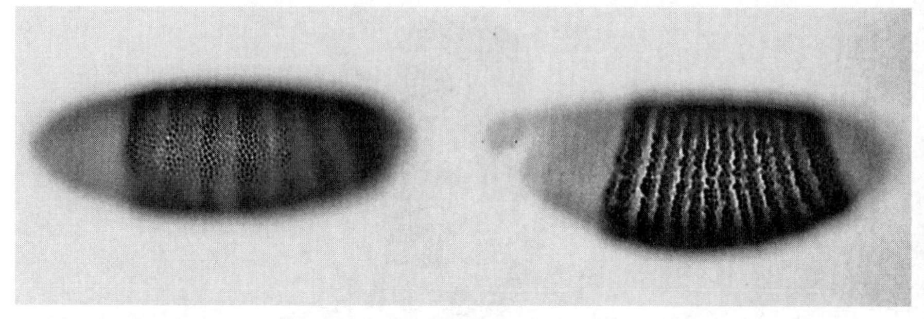

Fig. 5. The expression of the *fushi tarazu* and *even skipped* gene products at two times in the early development of the *Drosophila* embryo. Reproduced with permission from Lawrence (1992, Plate 4.1).

the bithorax phenotype, for example, block some downstream genes from being transcribed in the normal manner during development. The products of those downstream genes are required to specify the normal identity of the third thoracic segment.

At least a score of hox genes play crucial roles in laying down segment boundaries and establishing segment identities in the early *Drosophila* embryo. Figure 5 shows staining for the products of just two hox genes at two very early stages in the development of the *Drosophila* embryo. As shown, the formation of boundaries is quite dramatic. Figure 6 shows a particularly elegant relationship of seven of the hox genes in *Drosophila*. To a first approximation, the expression of each gene is required not only to help establish segment boundaries but also to activate the next gene in the series. That product is required to block the expression of the preceding gene, to delimit and help specify the identity of the next segment, and to activate the genes required for the succeeding segment. This apparently explains the fact that the physical position of these genes on the chromosome is highly conserved and corresponds with the geographical position of the segments in the embryo in which they are expressed. The second approximation takes into account the controls that affect the timing of gene expression and the fact that the proteins involved interact with each other and with the machinery for transcribing genes in establishing boundaries; fortunately, we do not need to go into the messy details for present purposes.

Let me connect these findings to my claims about homology at the beginning of this section and in the previous chapter. I claimed that distinct processes can trigger formation of homologous segments. This is true for formation of a second thoracic segment as the *bithorax* mutation shows. (Roughly speaking, formation of the first copy of the second thoracic segment

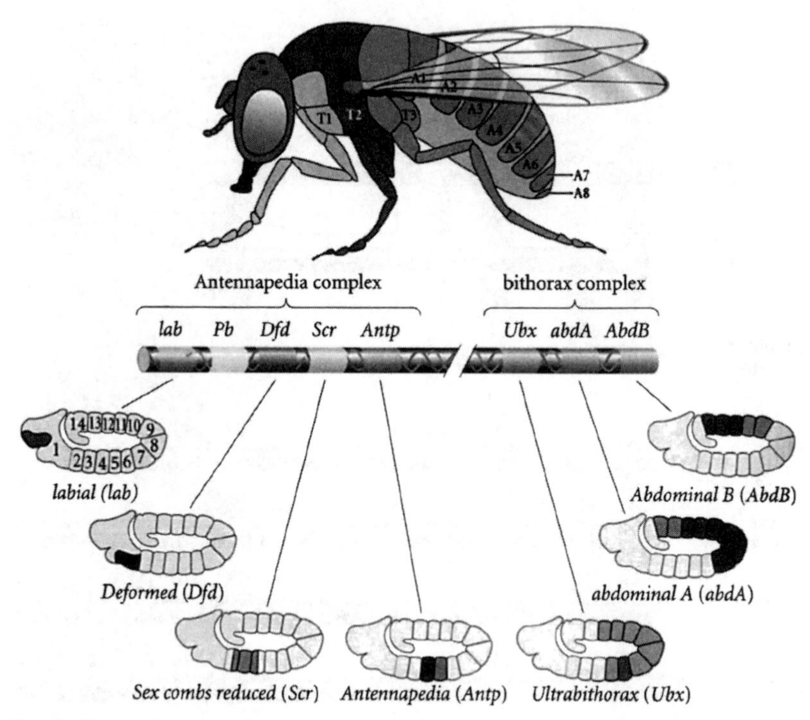


Fig. 6. Homeotic gene expression in *Drosophila*. The center bar represents two parts of chromosome three, with two disjoint regions containing the antennapedia and bithorax gene complexes. Above and below are the corresponding regions in which the products of these genes are expressed and in which mutations of the genes yield homeotic changes. Above is the adult, below the blastoderm of the embryo. Reproduced with permission of Sinauer Associates and Prof. Gilbert (2003, Fig. 9.28), after Dessain et al. (1992) and Kaufman, Seeger, and Olsen (1990).

is triggered by expression of the *ultrabithorax* gene, and formation of the second copy is triggered by preventing the ultrabithorax protein from triggering expression of the *abdominal a* gene.) The point is even clearer from a comparison of grasshoppers and *Drosophila*. Grasshoppers first make their cells and then, afterward, sequentially establish the segment identities for groups of cells, by means of (among other things) cell-to-cell signaling. *Drosophila* define their body segments before they make any cell walls – when the nuclei of the *Drosophila* cells are in a common cytoplasm, in which the boundaries are laid down *before* the cells are formed. So, cell-to-cell signaling, crucial in grasshopper segmentation, plays no role at all in establishing *Drosophila* segments. Thus, grasshoppers build segments by a different process than *Drosophila*, even though the same hox genes and gene products determine

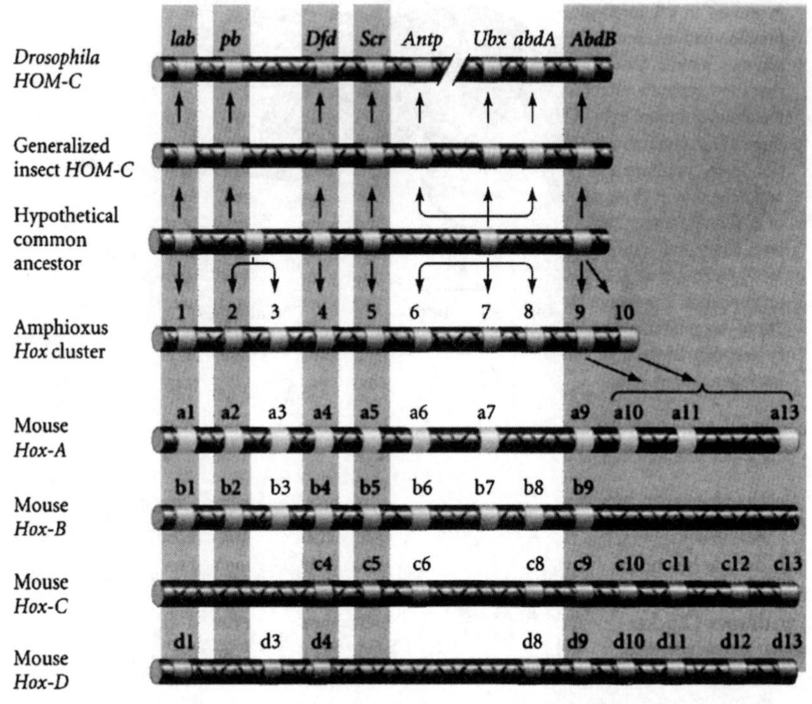


Fig. 7. Postulated ancestry of the homeotic genes from a hypothetical distant ancestor of both insects and mammals. Amphioxus, considered a direct ancestor of vertebrates, still has only one chromosome with these homeotic genes. Mammals have four such chromosomes. Reproduced with permission of Sinauer Associates and Prof. Gilbert (2003, Fig. 23.11), after Holland and Garcia-Fernández (1996).

the homologous segment boundaries. Conclusion: *Homologous segments can be formed by distinct processes.*

In *Drosophila*, eight hox genes are found on two separate regions of chromosome 3; however, in some insects and primitive chordates they are closely linked on one chromosome (Figure 7). Sometime in chordate evolution, in the lineage leading to vertebrates, the chromosome carrying those genes doubled two or three times, eventually yielding four chromosomes, on some of which there were local doublings that resulted in a few extra hox genes. Thus, *all* vertebrates have at least four clusters of hox genes (though none is complete). But, it is still possible to identify safely which vertebrate hox genes are homologous with which *Drosophila* hox genes. The hox genes are all laid out in the same order as in *Drosophila* on all four vertebrate chromosomes, although with some genes omitted on each chromosome. All these genes are activated at some time in development. In some cases, if one of them is

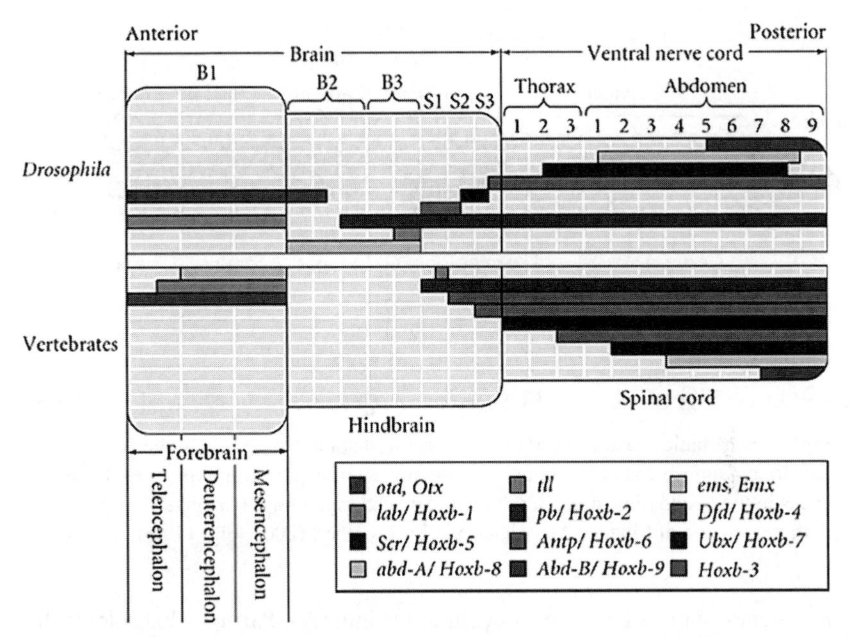


Fig. 8. Expression of regulatory transcription factors in *Drosophila* and in vertebrates along the anterior–posterior axis. Reproduced with permission of Sinauer Associates and Prof. Gilbert (2003, Fig. 23.2), after Hirth and Reichert (1999).

mutated or deleted in the laboratory, another one, usually on one of the other chromosomes, can perform some or all of its functions.

The findings about the functions of the control modules that regulate production of the products of these homologous genes yield a startling conclusion. In comparing the expression of the segment-defining genes we saw in *Drosophila* with the generalized picture of the expression of the homologous genes in vertebrates (see Figure 8), we see that – considered at an appropriate level – strongly homologous control systems are at play here. Without going into further detail, one can see that there is genuine segmentation in vertebrates and that homologous controls determine the segment identities of insects and vertebrates. Yet, there is no risk of confusing the segments of a chicken, a giraffe, a mouse, or a human with each other or with those of a fruit fly. The controls that determine segment boundaries and segment identities yield quite different products. In short, *homologous processes can yield utterly distinct products*.

I take two more small steps before turning again to the controls for making eyes introduced in the previous chapter. First, a slightly finer-grained example reinforces the idea that tinkering with the control systems utilizing these

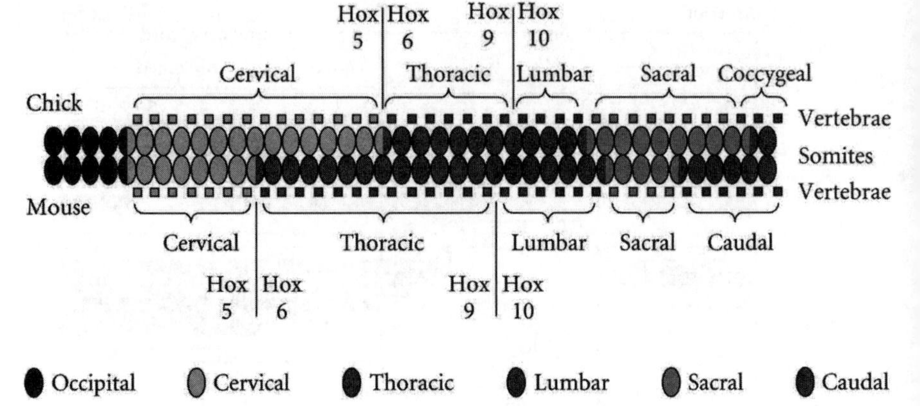


Fig. 9. Schematic illustration of the expression of some Hox genes in the vertebrae of developing mouse and chick embryos along the anterior-posterior axis. Note the differences in the boundaries of the paralogous (i.e., corresponding) Hox genes. Reproduced with permission of Sinauer Associates and Prof. Gilbert (2003, Fig. 11.46) after Burke et al. (1995).

hox genes plays a key role in evolution (Figure 9). Part of what allows the formation of the extra cervical vertebrae formed during ontogeny of chickens is the shifting of the position at which *Hox6* is expressed farther toward the end of the organism. Similarly, to produce specialized thoracic vertebrae like those required for wings in just the right region of the body, the *Hox9/Hox10* expression boundary is shifted backward and the boundary between thoracic and lumbar slightly forward. Although we have not examined how to change the adult structures produced at a given segment and the control system is baroquely complicated in ways I have not described, we can see that (and glimpse how) altering segment identities facilitates evolutionary change. Furthermore, an account rather similar to the one I have given for the axis from front to back also applies, although with considerable differences of detail, to the proximal to distal axis in segments and limbs.[8] This means that the tinkering idea looks plausible there as well. Thus, the account I have offered for assigning identities along the major body axis by use of control modules may be extended, with modifications, to cover the assignment of identities on other axes and to many body compartments, and is likely to be central to an account of a great deal of development. Repetition of similar controls at many levels and in novel combinations contributes to the regulation of gene expression and helps explain how particular gene products are made only in

[8] For example, a homeotic mutation in insects that replaces an antenna with a leg orders the parts of the leg, proximal to distal, correctly.

certain biochemical milieux, in certain compartments of the body, and, in some cases, at certain stages of development.

We are ready to take another look at some of the recent work on the switches that turn on the process of making eyes. Within the last decade, various labs have shown that activation of so-called selector genes can turn on the elaborate cascade of events to make certain organs in *Drosophila*. This was first shown for eyes (Gehring 1998; Halder, Callaerts, and Gehring 1995; see Chapter 11) and recently for wings (Guss et al. 2001). The story is interesting, and it depends on a variety of biotechnological tricks. The combination of tricks allows one to express the gene of interest virtually at pleasure by using a normally irrelevant signal to turn on the gene in a specific patch of tissue to learn what happens when it is expressed in the "wrong" place or at the "wrong" time.

In the laboratory of Walter Gehring in Basel, this was done in 1995 with a gene originally named *eyeless* in *Drosophila* because, when mutated, it yielded a fly without eyes. When the unmutated version of the eyeless gene was activated in various tissues, supernumerary eyes were produced (see Chapter 11, Figure 5). But, the situation proved to be yet more interesting – and raised a problem that will occupy us for years. The sequence of the eyeless gene showed it to be a hox gene. So, the people in the Gehring lab searched in the sequence databases for apparently homologous hox genes and they found them to be ubiquitous in animals. Indeed, the follow-up has shown that the genes *are* homologous (because their extraordinary similarity is due to being derived from a common source) and that they play a crucial role in eye formation in almost anything that has eyespots or eyes.[9] (The homologous gene is also expressed in the anterior parts of the heads of primitive animals that have no eyes, so it probably was not originally involved in making eyes.)

In 1995, the mouse gene had recently been renamed *Pax6*.[10] In an already classic experiment, Gehring's group employed the mouse gene just as they had employed the *eyeless* gene in *Drosophila* – that is, they activated it in various tissues of a developing *Drosophila*. Figure 6 in Chapter 11 shows what

[9] There is a clear consensus that the circuitry of the regulatory module has been reasonably well elucidated. For example, Pineda et al. (2000) have shown that the same regulatory circuit, or process module, drawing on homologous genes and gene products, deployed in the same order as in vertebrates and insects, is present in *platyhelminths* (i.e., flatworms), where it is required for the formation of eyes. They argue in considerable molecular detail that the regulatory circuit is evolutionarily conserved and is found throughout triploblastic animals (i.e., animals with three-layered embryos, which include flatworms, arthropods, and vertebrates).

[10] As indicated in the previous chapter, it was originally known as *aniridia* in mice because, when mutated, it yielded mice whose eyes had no iris; and as *small eye* in humans because that was the phenotype produced when the gene mutated.

they got. Expression of the mouse *Pax6* gene in *Drosophila* acted just like expression of *eyeless* (the *Drosophila Pax6* gene), switching on the cascade of events that make a *Drosophila* eye. Recently, the *Drosophila Pax6* gene was used to make ectopic eyes in the frog *Xenopus* (Chow et al. 1999).[11] Arguably, these results show that *homologous processes can trigger the formation of organs that perform similar functions, but – because the structures produced have many features that are not shared and not derived from a common source – are not closely homologous.* Of course, this result has many other consequences and raises many more questions. I return to some of them in the concluding discussion.

Before concluding, I turn to one last figure to make a final point (Figure 9). Gehring's group described the *Pax6* gene as a master control gene for making eyes. I think that description is seriously misleading. The gene itself is not a control device; its expression is tightly controlled. Following are four arguments for this claim:

1. As the schematic bar in Figure 10 shows, in mice, the gene is expressed in at least four different places, one of which is the pancreas. When the gene is expressed in the pancreas, it does not make eyes. Rather, it and its gene product interact with other regulatory genes and proteins to regulate development of pancreatic cells and transcription of the insulin, glucagon, and somatostatin genes of the pancreas (Andersen et al. 1999; Hussain and Habener 1999; Sander et al. 1997, reviewed in Gilbert 2003, pp. 116–17).
2. What switches on the cascade that makes the eye or that produces pancreatic proteins is a group of interacting signal-transduction modules hooked into the right context. The multiple modules required to initiate the different processes are composed not only of genes but also of their own gene products and a series of additional proteins that must interact correctly, with each other and with the nucleotide sequences that respond to the signals. Thus, neither a gene by itself nor its products by themselves are controllers. The genes regulated by *Pax6* will not initiate the eye-making cascade or production of pancreatic proteins unless they are operating within the context of the relevantly related groups of signal-transduction

[11] However, the *Drosophila* gene appears not to produce ectopic eyes in mice. According to Gilbert (2003, p. 120), this is because the *Drosophila Pax6* gene product does not repress the mouse genes that are used to construct the organs that would normally be produced in locations downstream from eyes, whereas the mouse *Pax6* gene product does repress the downstream triggers for formation of "downstream" organs in *Drosophila*.

(A)

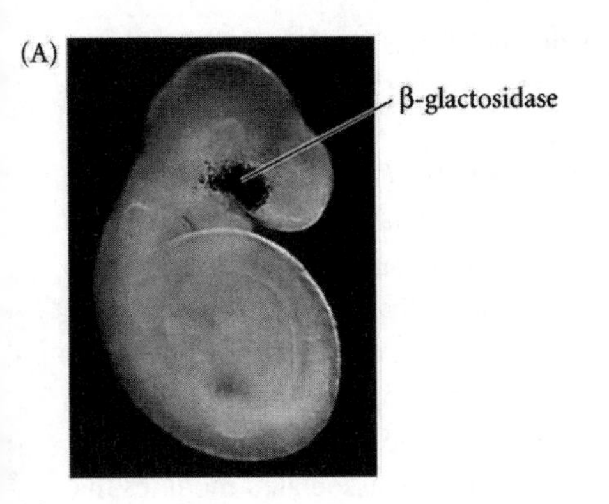


(B)

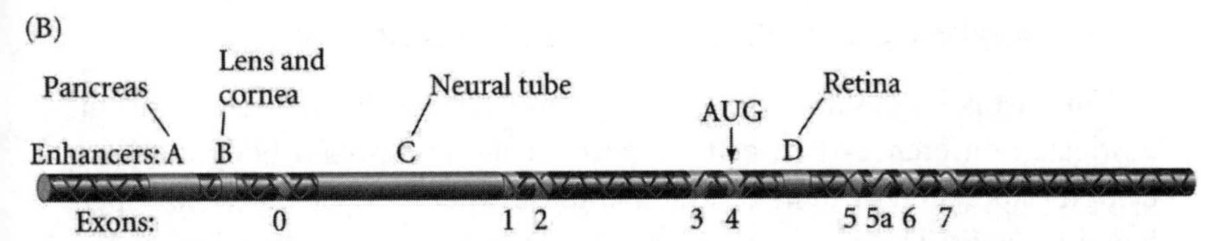


Fig. 10. Expression of a mouse *Pax6* gene when the lens and cornea enhancer is activated in embryogenesis. The gene was altered by biotechnology so that its product could be visualized. The schematic bar shows the mouse *Pax6* gene with its regulatory regions. Reproduced with permission of Sinauer Associates and Prof. Gilbert (2003, Fig. 5.7). After Williams et al. (1998) and Kammandel et al. (1998).

modules. The result of signal transduction depends on what signal is transduced and on the cellular and biochemical context into which the signal transducing module is introduced. We have seen that the particular module involving *Pax6* serves to switch on the eye-making cascade only in some cellular and subcellular contexts, not in others, so that expression of *Pax6* does not universally switch on the eye-making apparatus. *Pax6* can be used in a variety of ways, depending on how its expression is regulated and what context its product is put into when that product is expressed. Recently, many biologists and philosophers have made related claims about genes in general (e.g., Griffiths and Neumann-Held 1999; Keller 2000, 2003; Morange 2001; Moss 2001, 2003; Neumann-Held 2001).

3. Even when genes or their direct products do perform sharply defined functions – for example, switching on the process of making eyes – they may not be the only genes that can perform those functions. Consider the impact of the sorts of gene duplication we saw with the hox genes. In the

case of eye-making, at least four distinct genes in *Drosophila* are now known whose products can turn on the cascade of events that normally result in an eye (Czerny et al. 1999).

4. *The eye-making cascade, even when initiated by Pax6, is tightly controlled by a higher order system. For this reason, the metaphor of a master control gene is misplaced.* The extraordinary evolutionary conservation of *Pax6* and various features of the regulatory network that initiates eye formation is not evidence for the specific power of the gene or its product. Since that gene and the associated control apparatus are coopted as control elements in many other cascades, control of what the gene does resides elsewhere. The extraordinary conservation of *Pax6* and the associated modules that control eye formation shows the power of the regulatory apparatus that has stabilized and regulated the means of producing eyes.

This last point deserves further discussion. Kango-Singh et al. (2003) investigated differences between the parts of the *Drosophila* body at which ectopic expression of *eyeless* does and those where it does not yield supernumerary eyes. The authors show that *eyeless* expression will not initiate eye formation except in the present of two secretory molecules (both products of regulatory genes), *hedgehog* and *decapentaplegic*. The authors propose that the further restrictions on where *eyeless* is able to trigger eye formation suggest that *eyeless* requires active collaboration of five additional regulatory genes to initiate the eye-making cascade.[12] I interpret this work as uncovering some of the mechanisms by means of which the eye-making cascade is controlled and restricted. I suggest that, in order for the control mechanisms to function, they require a reliable selector gene (or a controllable module that acts as a cascade initiator) to produce a stable functional eye. The selector gene is thus a critical cog in the control system, but is not the master control device. This is not surprising: initiation of the complex cascade that yields eyes (and thus of the genes that initiate it) *has* to be rigorously controlled – otherwise there would be organisms with eyes in all sorts of surprising places. What switches on the cascade in normal circumstances is not just the expression of the *Pax6* gene but an integrated module or group of interacting modules, built out of numerous interacting transducing signals set into the appropriate contexts at various stages of development.

Gehring and his colleagues are not convinced by such arguments as these and continue to argue strongly for their view that "*Pax6* is the universal master

[12] The details are rather complex and beyond the scope of the present investigation. The genes in question are called *eyes absent, sine oculis, dachshund, eyes gone, and twin of eyeless* (Kango-Singh et al. 2003, 58).

control gene for eye morphogenesis in metazoa ranging from platyhelminths to humans" (Gehring and Ikeo 1999, p. 376). They initially thought the *eyeless* [*Drosophila Pax6*] gene to be unique among the genes that initiate the eye-making cascade in being expressed before ("upstream of") the other genes that can initiate that cascade (Halder et al. 1995, p. 1788). But they quickly recognized that this is incorrect (Czerny, et al. 1999). Upstream expression of a gene named "twin of eyeless" (abbreviated "*Toy*") turns out to initiate eye formation in *Drosophila* by activating expression of *eyeless* and a number of additional genes that also can initiate the eye-making cascade in *Drosophila* (as can some of their homologues in some other organisms). In spite of these and other qualifications of their initial claims, Gehring and others argue strenuously that the description of *Pax6* is appropriate. Gehring and Ikeo (1999), for example, provide additional evidence, beyond the scope of this discussion, for their position. Their review covers many of the points just made (and many more), but interprets them as supporting their claims about the status of *eyeless* [*Pax6*]. Since these are true experts in this domain, I advance my contrary views with some diffidence. But what is at stake are modularly constructed networks and canonical pathways whose complex interactions are stabilized after millions of generations. Whether the dramatic findings of the Gehring's laboratory (and many others) justify claims to have found a master control gene will remain a topic for debate for many years to come.

Thus, there is no determinate answer to what this gene does – or, more generally, what any gene does. What genes do typically depends on when and where they are expressed and how they are spliced, and this depends on the system of controls that activates or represses them and regulates the splicing. Until one knows in what contexts they are expressed, how they are spliced in those contexts, how their products are post-transcriptionally or post-translationally modified, and what processes they enter into in those contexts, one's account is deficient. In the life of an individual organism, the answers to these questions depend on a great variety of contingencies, including which other genes are nearby and (because regulatory systems respond to feedback) which biochemical and other modifications have already occurred in response to the environment that the organism and the relevant tissues have already encountered. Thus, the regulatory controls are at least partly epigenetic.

These arguments are quite general. They show (what Morgan et al. (1915) already knew about genes in 1915, though not sequences!) that the salient effects by means of which we identify Mendelian genes need not be the most important effects of the sequences we pick out by reference to the associated phenotype. They also show that the precise effects of genes interpreted as sequences cannot be determined by their sequences or even the sequence of the entire genome. To put it strongly: Without knowing an enormous amount about the contexts in which a sequence and its products are placed – contexts that vary enormously in ways that cannot be predicted from complete knowledge of the genome – there is, in general, no determinate answer to the question, "What does this sequence of nucleotides do for this organism?" Thus, if one tries to proceed strictly from the genome up, one cannot predict, in general, what effects sequence-identified genes will have on the organism and how they will affect its fitness. At best, we can obtain contextually restricted propensities or ex post facto averages.

## PHILOSOPHICAL CODA

To close, I first summarize the major points for which I have provided a direct argument and then turn to some larger speculations. I believe I have made a persuasive case for the following claims:

1. Signal-transduction processes commonly regulate the contexts in which many proteins are produced and the amino-acid sequences of the proteins[13] and (indirectly thereby) what those proteins do. The signal-transduction pathways are themselves multipurpose modules that have been recycled through evolutionary time into many different contexts. Because a given sequence of amino acids can take on radically different roles in different tissues and at different stages in ontogeny, the pathway from genotype to phenotype is not determined by nucleotide sequence.
2. Because the pathways from genes to proteins are far more complex than was traditionally expected, the pathway from a gene to the totality of phenotypic traits it causes or affects is even more difficult to analyze. In general, if genes are identified with nucleotide sequences, it is not possible to determine in any straightforward way the totality of effects of a gene on the particular animal within which it occurs.
3. Genes, taken by themselves, are not central agents in any of the processes considered. Activation and control of transcription depends thoroughly on complex systems for making, activating, and regulating the action of the enzymes that produce RNA transcripts. This complex process is not controlled by the genes that are transcribed – although it is often modulated by signals that are, themselves, sequences of nucleotides or derived from sequences of nucleotides.

[13] My direct argument turns on the regulation of DNA transcription, alternative splicing, and intranuclear processing of RNA. Other controls of post-transcriptional and post-translational processing, not discussed here, only strengthen the argument.

Given these and related considerations, no serious version of genetic determinism can be true. The sequence of nucleotides in the genome of an egg, sperm, or fertilized egg is not a blueprint for the organism.

These considerations are closely allied to the view that evolution is a process of tinkering. We can now see that animals are composites in surprising ways. Their ontogeny is controlled by a miscellany of processes that come from different sources and work in different ways. To appreciate the full depth and strength of this point, one has to go far beyond the material covered in this chapter. Evolutionary tinkering means that items from different sources have been put together in novel, often awkward ways.

A related phenomenon has some bearing on this issue. Many cascades triggered by the activation of particular modules, among them the formation of an "extra" second thoracic segment or a supernumerary eye, exhibit considerable autonomy. Many of the processes we have considered (though not some of the earliest ones involved in segmentation and specification of segment identities) are surprisingly independent of each other once they begin – formation of the primary body axis, formation of limb axes, formation of a specific organ (not just the eye – also a wing, a bone, a limb, a heart, a kidney, or a brain). Other things being equal, each of these is carried out to completion, whatever the fate of the others. Similar things are true for process on many scales that have not been considered herein. Two examples, chosen simply to illustrate that similar claims hold at different levels, follow: (1) construction and assembly of the intracellular microtubule spindle apparatus required for mitosis, and (2) making the branches of nerves in the nervous system. All of the processes named are under *local* control, which is how they can carry on independently of other processes on the same scale. The signals that limit an organ or that limit limb-bud growth, that stop membrane growth, microtubule extension, and cell proliferation – all seem to be based on short-range signals or local biochemistry. These considerations suggest that there are numerous quasi-autonomous processes that are somehow coordinated in making a complex multicellular animal.

The quasi-independence of processes just suggested makes good evolutionary sense. In effect, for evolution to occur, traits must be capable of varying independently. The modules that control those traits, therefore, must be mutually independent. After all, if forelimbs could not vary independently from hindlimbs, a single lineage could not produce opossums, koalas, and kangaroos. Quasi-autonomy at many levels allows tinkering; however, quasi-autonomy of this sort means that there cannot be too close a connection between the controls that regulate one part and the controls that regulate other parts, even closely related parts. Speculatively, I suggest that distant signals (e.g., the signals of circulating hormones) coordinate fairly independent cascades of events that are "locally" controlled by particular modules or module cascades that draw on immediately available resources – including, of course, the nucleotide sequences of the genes inside the relevant cells. The idea here is that coordination of ontogeny is achieved by dependence on or integration of a cascade of distant signals that allows the start of a process that depends on distant assembly processes having already been completed.

However, this does not require that there be any "master plan" in virtue of which the body is constructed. Rather, the integration of the body is the evolutionary result of quasi-autonomous modules locking in and starting irreversible processes in ways that are coordinated by distant signals. This coordination is under the long-term but rigorous constraints imposed by natural selection, which usually allows flexibility in the details of loosely linked developmental processes, but culls those organisms that are not as well integrated as their closest competitors. The timing and location of major processes (segmentation, establishing boundaries between types of segments, organ, and limb formation, etc.), when varied, can yield novel results because of changed genetic and epigenetic background. In such cases, the modification of quasi-autonomous processes may yield alterations of organs, limbs, and, rarely, evolutionary novelties and change in *Bauplan* (Newman and Müller 2000; Müller and Newman 2003). In short, coordination of the shuffling of quasi-autonomous processes by use of long-range signals provides a possible mechanism for two major classes of developmental change in evolutionary time. The first class is the shuffling of relationships among the evolutionary stable products of those processes (such as changes in allometry and in relative capacities of organs). The second class can occur when processes are introduced into sufficiently novel genetic or epigenetic backgrounds that their products or the relationships among them are transformed sufficiently to yield evolutionary novelties.

If this approach withstands criticism, it will lend considerable support for one of the major claims put forward in this chapter: *the key to the integration of organisms is not dependent on a master plan, but on the coordination of quasi-autonomous modules, which are not controlled primarily – let alone solely – by genes or the genetic material.*

## REFERENCES

Andersen, F. G., R. S. Heller, H. V. Petersen, J. Jensen, O. D. Madsen, and P. Serup. 1999. "Pax6 and Cdx2/3 form a functional complex on the rat glucagon gene promoter G1-element." *FEBS Letters* 445: 306–10.

Arthur, W. 1988. *A Theory of the Evolution of Development*. Chichester: Wiley.
Arthur, W. 1997. *The Origin of Animal Body Plans: A Study in Evolutionary Developmental Biology*. Cambridge: Cambridge University Press.
Black, D. L. 1998. "Splicing in the inner ear: A familiar tune, but what are the instruments." *Neuron* 20: 165–8.
Breitbart, R. A., A. Andreadis, and B. Nadal-Ginard. 1987. "Alternative splicing: A ubiquitous mechanism for the generation of multiple protein isoforms from single genes." *Annual Review of Biochemistry* 56: 481–95.
Burke, A. C., A. C. Nelson, B. A. Morgan, and C. J. Tabin. 1995. "Hox genes and the evolution of vertebrate axial morphology." *Development* 121: 333–46.
Carroll, S. B., J. K. Grenier, and S. D. Weatherbee. 2001. *From DNA to Diversity: Molecular Genetics and the Evolution of Animal Design*. Malden, MA: Blackwell Science.
Chow, R. L., C. R. Altmann, R. A. Lang, and A. Hemmati-Brivanlou. 1999. "*Pax*6 induces ectopic eyes in a vertebrate." *Development* 126: 4213–22.
Czerny, T., G. Halder, U. Kloter, A. Souabni, W. J. Gehring, and M. Busslinger. 1999. "Twin of eyeless, a second Pax-6 gene of *Drosophila*, acts upstream of eyeless in the control of eye development." *Molecular Cell* 3: 297–307.
Davidson, E. H. 2001. *Genomic Regulatory Systems: Development and Evolution*. San Diego: Academic Press.
Dessain, S., C. T. Gross, M. A. Kuziora, and W. McGinnis. 1992. "*Antp*-type homeodomains have distinct DNA-binding specificities that correlate with their different regulatory functions in embryos." *EMBO Journal* 11: 991–1002.
Duboule, D., and A. S. Wilkins. 1998. "The evolution of 'bricolage'." *Trends in Genetics* 14, no. 2: 54–9.
Gehring, W. J. 1998. *Master Control Genes in Development and Evolution: The Homeobox Story*. New Haven: Yale University Press.
Gehring, W. J., and K. Ikeo. 1999. "Pax6: Mastering eye morphogenesis and eye evolution." *Trends in Genetics* 15: 371–7.
Gerhart, J., and M. Kirschner. 1997. *Cell, Embryos, and Evolution: Toward a Cellular and Developmental Understanding of Phenotypic Variation and Evolutionary Adaptability*. Malden, MA: Blackwell Science.
Gilbert, S. F. 2000. *Developmental Biology*. 6th ed. Sunderland, MA: Sinauer Associates.
Gilbert, S. F. 2003. *Developmental Biology*. 7th ed. Sunderland, MA: Sinauer Associates.
Griffiths, P. E., and E. M. Neumann-Held. 1999. "The many faces of the gene." *BioScience* 49: 656–62.
Guss, K. A., C. E. Nelson, A. Hudson, M. E. Kraus, and S. B. Carroll. 2001. "Control of a genetic regulatory network by a selector gene." *Science* 292: 1164–7.
Halder, G., P. Callaerts, and W. J. Gehring. 1995. "Induction of ectopic eyes by targeted expression of the eyeless gene in *Drosophila*." *Science* 267, no. 5205: 1788–92.
Hall, B. K. 1992. *Evolutionary Developmental Biology*. London: Chapman and Hall.
Hall, B. K., ed. 1994. *Homology: The Hierarchical Basis of Comparative Anatomy*. New York and San Diego: Academic Press.
Hall, B. K. 1999. *Evolutionary Developmental Biology*. Dordrecht, Holland: Kluwer.
Hall, B. K., and W. M. Olson. 2003. *Keywords & Concepts in Evolutionary Developmental Biology*. Cambridge, MA: Harvard University Press.

Hirth, F., and H. Reichert. 1999. "Conserved genetic programs in insect and mammalian brain development." *BioEssays* 21: 677–84.

Hogenesch, J. B., K. A. Ching, S. Batalov, A. I. Su, J. R. Walker, Y. Zhou, S. A. Kay, P. G. Schultz, and M. P. Cooke. 2001. "A comparison of the Celera and Ensembl predicted gene sets reveals little overlap in novel genes." *Cell* 106: 413–15.

Holland, P. W. H., and J. Garcia-Fernàndez. 1996. "Hox genes and chordate evolution." *Developmental Biology* 173: 382–95.

Horowitz, D. S., and A. R. Kraianer. 1995. "Mechanisms for selecting 5' splice sites in mammalian pre-mRNA splicing." *Trends in Genetics* 10: 100–6.

Hussain, M. A., and J. Habener. 1999. "Glucagon gene transcription activation mediated by synergistic interactions of *Pax*-6 and *Cdx*-2 with the p300 co-activator." *Journal of Biological Chemistry* 274, no. 41: 28950–7.

International Genome Sequencing Consortium. 2001. "Initial sequencing and analysis of the human genome." *Nature* 409: 860–941.

Jablonka, E., and M. J. Lamb. 1995. *Epigenetic Inheritance and Evolution: The Lamarckian Dimension*. Oxford and New York: Oxford University Press.

Jacob, F. 1977. "Evolution and tinkering." *Science* 196: 1161–6.

Jacob, F. 1982. *The Possible and the Actual*. Seattle: University of Washington Press.

Kammandel, B., A. Chowdhury, A. Stoykova, S. Aparicio, S. Brenner, and P. Gruss. 1998. "Distinct *cis*-essential modules direct the time-space pattern of *Pax6* gene activity." *Developmental Biology* 205: 79–97.

Kango-Singh, M., A. Singh, and Y. H. Sun 2003. "Eyeless collaborates with Hedgehog and Decapentaplegic signaling in *Drosophila* eye induction." *Developmental Biology* 256: 49–61.

Kaufman, T. C., M. A. Seeger, and G. Olsen. 1990. "Molecular and genetic organization of the Antennapedia gene complex of *Drosophila melanogaster*." *Advances in Genetics* 27: 309–62.

Keller, E. F. 2000. *The Century of the Gene*. Cambridge, MA: Harvard University Press.

Keller, E. F. 2003. *Making Sense of Life*. Cambridge, MA: Harvard University Press.

Laubichler, M. D. 2000. "The organism is dead. Long live the organism!" *Perspectives on Science* 8: 286–315.

Lawrence, P. A. 1992. *The Making of a Fly: The Genetics of Animal Design*. Oxford: Blackwell Scientific Publications.

Morange, M. 2001. *The Misunderstood Gene*. Transl. M. Cobb. Cambridge, MA: Harvard University Press.

Morgan, T. H., A. H. Sturtevant, H. J. Muller, and C. B. Bridges. 1915. The *Mechanism & Mendelian Heredity*. New York: Henry Holt and Co.

Moss, L. 2001. "Deconstructing the gene and reconstructing molecular developmental systems." In *Cycles of Contingency: Developmental Systems and Evolution*, eds. S. Oyama, P. E. Griffith, and R. D. Gray. Cambridge, MA: MIT Press, 85–97.

Moss, L. 2003. *What Genes Can't Do*. Cambridge, MA: MIT.

Müller, G. B., and S. A. Newman, eds. 2003. *Origination of Organismal Form: Beyond the Gene in Developmental and Evolutionary Biology*. Cambridge, MA: MIT Press.

Neumann-Held, E. M. 2001. "Let's talk about genes: The process molecular gene concept and its context." In *Cycles of Contingency: Developmental Systems and Evolution*, eds. S. Oyama, P. E. Griffith, and R. D. Gray. Cambridge, MA: MIT Press, 69–84.

Newman, S. A., and G. B. Müller. 2000. "Epigenetic mechanisms of character origination." *Journal of Experimental Zoology (Molecular and Developmental Evolution)* 288: 304–17.

Pigliucci, M. 2001. *Phenotypic Plasticity: Beyond Nature and Nurture*. Baltimore: Johns Hopkins University Press.

Pineda, D., J. Gonzalez, P. Callaerts, K. Ikeo, W. J. Gehring, and E. Salo. 2000. "Searching for the prototypic eye genetic network: *Sine oculis* is essential for eye regeneration in planarians." *Proceedings of the National Academy of Sciences, USA* 97: 4525–9.

Raff, R. A. 1996. *The Shape of Life: Genes, Development, and the Evolution of Animal Form*. Chicago: University of Chicago Press.

Raff, R. A., and T. C. Kaufman. 1983. *Embryos, Genes, and Evolution: The Developmental-Genetic Basis of Evolutionary Change*. New York: Macmillan.

Sander, M., A. Neubuser, J. Kalamaras, H. C. Ee, G. R. Martin, and M. S. German. 1997. "Genetic analysis reveals that Pax6 is required for normal transcription of pancreatic hormone genes and islet development." *Genes and Development* 11: 1662–73.

Schlichting, C. D., and M. Pigliucci. 1998. *Phenotypic Evolution: A Reaction Norm Perspective*. Sunderland, MA: Sinauer Associates.

Venter, J. C., et al. 2001. "The sequence of the human genome." *Science* 291: 1304–51.

Wagner, G. P., ed. 2001. *The Character Concept in Evolutionary Biology*. San Diego: Academic Press.

Wilkins, A. S. 2002. *The Evolution of Developmental Pathways*. Sunderland, MA: Sinauer Associates.

Williams, S. C., C. R. Altmann, R. L. Chow, A. Hemmati-Brivanlou, and R. A. Lang. 1998. "A highly conserved lens transcriptional control element from the *Pax-6* gene." *Mechanics of Development* 73: 225–9.

Wright, F. A., W. J. Lemon, W. D. Zhao, R. Sears, D. Zhuo, J. P. Wang, H. Y. Yang, T. Baer, D. Stredney, J. Spitzner, et al. 2001. "A draft annotation and overview of the human genome." *Genome Biology* 2, no. 7: 1–18.

# Index